I0167933

Leader Development FM 6-22
Red Bike Publishing Huntsville, AL
www.redbikepublishing.com

Published in the United States of America

ISBN-13: 978-1-936800-28-5

*FM 6-22

Field Manual
No. 6-22

Headquarters
Department of the Army
Washington, DC, 30 June 2015

Leader Development

Contents

Distribution Restriction: Approved for public release; distribution unlimited.

*This publication supersedes FM 6-22, dated 12 October 2006.

Figures

Tables

Preface

FM 6-22 *Leader Development* provides a doctrinal framework covering methods for leaders to develop other leaders, improve their organizations, build teams, and develop themselves.

The principal audience for FM 6-22 is all leaders, military and civilian, with an application focus at the operational and tactical levels. Trainers and educators throughout the Army will also use this manual.

Commanders, staffs, and subordinates ensure their decisions and actions comply with applicable U.S., international, and, in some cases, host-nation laws and regulations. Commanders at all levels ensure their Soldiers operate in accordance with the law of war and the rules of engagement (see FM 27-10).

FM 6-22 uses joint terms where applicable. Selected joint and Army terms and definitions appear in both the glossary and the text. Terms for which FM 6-22 is the proponent publication (the authority) are marked with an asterisk (*) in the glossary. Definitions for which FM 6-22 is the proponent publication are boldfaced in the text. For other definitions shown in the text, the term is italicized and the number of the proponent publication follows the definition.

This publication incorporates copyrighted material.

FM 6-22 applies to the Active Army, Army National Guard/Army National Guard of the United States, and United States Army Reserve unless otherwise stated.

The proponent of FM 6-22 is Headquarters, U.S. Army Training and Doctrine Command. The preparing agency is the Center for Army Leadership, U.S. Army Combined Arms Center—Mission Command Center of Excellence. Send comments and recommendations on DA Form 2028 (*Recommended Changes to Publications and Blank Forms*) to Center for Army Leadership, ATTN: ATZL-MCV-R, 290 Stimson Avenue, Fort Leavenworth, KS 66027-1293; by e-mail to usarmy.leavenworth.tradoc.mbx.6-22@mail.mil; or submit an electronic DA Form 2028.

LIVING DOCTRINE

Bringing Doctrine to Life

The Army is committed to delivering doctrine to our Soldiers and civilians through the various media used in everyday life. FM 6-22 is available in an eReader format for download to commercial mobile devices from the Army Publishing Directorate (www.apd.army.mil). A platform-neutral application (LeaderMap) has also been developed to augment the content of the manual with additional multi-media material. LeaderMap is available thru the Central Army Registry (www.adtdl.army.mil) and can be found by typing LeaderMap into the search function after signing in. A fully enhanced interactive version of FM 6-22 for commercial devices will be available at the Army Training Network (https://atn.army.mil/). The fully enhanced interactive version integrates video, audio, and interactivity to enhance the overall learning and reading experience. An announcement will be made Armywide as soon as the interactive version is fielded.

Acknowledgements

This manual contains copyrighted material as indicated:

Chapter 2, paragraphs 2-6, 2-7, and 2-28 and the example scorecard; Chapter 3, paragraphs 3-4–3-5, 3-7–3-9, 3-11–3-13,4, 3-20–3-34, 3-57–3-60, 3-106, 3-117–3-122, 3-124–3-126, 3-134–3-136, and special callout texts within these paragraphs; and chapter 4, paragraphs 4-57–4-59 and the personal after action review (AAR) on page 4-12 come from *Commander's Handbook for Unit Leader Development*, Copyright © 2007 United States Government, as represented by the Secretary of the Army. All rights reserved.

Chapter 4, paragraphs 4-8–4-15, 4-19–4-57, and 4-61–4-74 and the analysis exercises on pages 4-4 and 4-5 come from *Self-Development Handbook,* Copyright © 2008 United States Government, as represented by the Secretary of the Army. All rights reserved.

Introduction

Army leaders are the competitive advantage the Army possesses that technology cannot replace nor be substituted by advanced weaponry and platforms. Today's Army demands trained and ready units with agile, proficient leaders. Developing our leaders is integral to our institutional success today and tomorrow. It is an important investment to make for the future of the Army because it builds trust in relationships and units, prepares leaders for future uncertainty, and is critical to readiness and our Army's success. Leader development programs must recognize, produce, and reward leaders who are inquisitive, creative, adaptable, and capable of exercising mission command. Leaders exhibit commitment to developing subordinates through execution of their professional responsibility to teach, counsel, coach, and mentor subordinates. Successful, robust leader development programs incorporate accountability, engagement, and commitment; create agile and competent leaders; produce stronger organizations and teams; and increase expertise by reducing gaps between knowledge and resources.

Leader development involves multiple practices that ensure people have the opportunities to fulfill their goals and that the Army has capable leaders in position and ready for the future. The practices include recruiting, accessions, training, education, assigning, promoting, broadening, and retaining the best leaders, while challenging them over time with greater responsibility, authority, and accountability. Army leaders assume progressively broader responsibilities across direct, organizational, and strategic levels of leadership.

FM 6-22 integrates doctrine, experience, and best practices by drawing upon applicable Army doctrine and regulations, input of successful Army commanders and noncommissioned officers, recent Army leadership studies, and research on effective practices from the private and public sectors.

FM 6-22 provides Army leaders with information on effective leader development methods by:

- Translating Army leader feedback into quick applications.
- Prioritizing leader development activities under conditions of limited resources.
- Integrating unit leader development into already occurring day-to-day activities.
- Integrating ADRP 6-22 leader attributes and competencies consistently across Army leader development doctrine.

FM 6-22 contains seven chapters that describe the Army's view on identifying and executing collective and individual leader development needs:

Chapter 1 discusses the tenets of Army leader development, the purpose of developing leaders to practice the mission command philosophy, building teams, and development transitions across organizational levels.

Chapter 2 discusses the creation of unit leader development programs.

Chapter 3 addresses the fundamentals for developing leaders in units by setting conditions, providing feedback, and enhancing learning while creating opportunities.

Chapter 4 provides information on the self-development process including strengths and developmental needs determination and goal setting.

Chapter 5 discusses character, judgment and problem solving, and adaptability as situational leader demands.

Chapter 6 provides information on leader performance indicators to enable observations and feedback.

Chapter 7 provides recommended learning and developmental activities.

The References section includes pertinent links to recommended leader development readings and Web sites.

Introductory figure 1 illustrates how the information within this manual fits together.

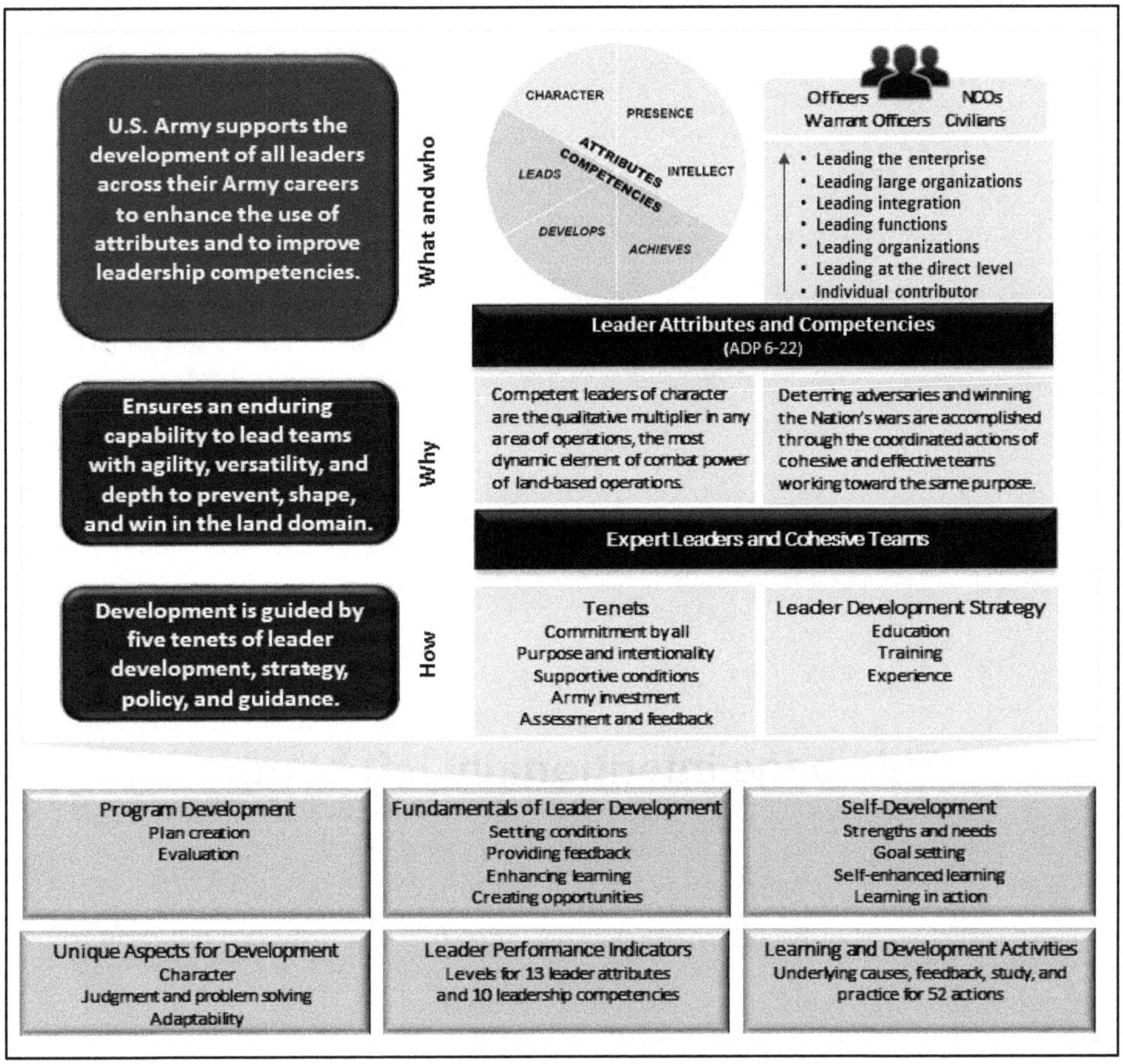

Introductory Figure 1. Integrating diagram

This page intentionally left blank.

Chapter 1
Leader Development

1-1. The Army depends upon itself to develop adaptable leaders able to achieve mission accomplishment in dynamic, unstable, and complex environments. A robust, holistic leader development program is essential. Through a mix of education, training, and experience, Army leader development processes produce and sustain agile, adaptive, and innovative leaders who act with boldness and initiative in dynamic, complex situations to execute missions according to doctrine, orders, and training. Furthermore, it also produces leaders that possess the integrity and willingness to act in the absence of orders, when existing orders, doctrine or their own experience no longer fit the situation, or when unforeseen opportunities or threats arise. Properly designed leader development programs develop trusted leaders of character, competence, and commitment. The goal is to develop Army leaders who clearly provide purpose, direction, motivation, and vision to their teams and subordinates while executing missions to support their commander's intent. Leaders at all levels need to be prepared to understand the strategic context for execution and success of any mission.

1-2. Leader development is fundamental to our Army—*leader development* is the deliberate, continuous, sequential, and progressive process—founded in Army values—that grows Soldiers and Army Civilians into competent and confident leaders capable of decisive action. Leader development is achieved through the life-long synthesis of the knowledge, skills, and experiences gained through the training and education opportunities in the institutional, operational, and self-development domains (AR 350-1). A key component of leader development is remaining focused on the professionalism of our leaders and those they lead. By developing and promoting a professional force, the Army develops trust on several levels: between Soldiers; between Soldiers and leaders; between Soldiers and Army Civilians; between the Soldiers, their families and the Army; and between the Army and the American people. This is why the Army is committed to providing quality institutions of education and training along with challenging experiences throughout a career.

TENETS OF ARMY LEADER DEVELOPMENT

1-3. The tenets of Army leader development provide the essential principles that have made the Army successful at developing its leaders. The tenets also provide a backdrop for the Army principles of unit training (see ADRP 7-0). The overarching tenets of Army leader development are—

- Strong commitment by the Army, superiors, and individuals to leader development.
- Clear purpose for what, when, and how to develop leadership.
- Supportive relationships and culture of learning.
- Three mutually supportive domains (institutional, operational, and self-development) that enable education, training, and experience.
- Providing, accepting, and acting upon candid assessment and feedback.

1-4. Development of people is an Army priority. Commitment represents intention and engagement from the individual, from supportive leaders, and from the Army. Beyond their directed responsibility to develop subordinates, leaders want to serve in an organization that values camaraderie and teamwork and improves the capabilities of others. Leaders have a directed responsibility to develop their subordinates; accountability for implementation follows responsibility. Leaders must be committed to the development of others and themselves. Teams change and organizations change when individuals choose to engage and improve.

1-5. Development depends on having clear purpose for what, when and how to develop. Good leader development is purposeful and goal-oriented. A clearly established purpose enables leaders to guide, assess, and accomplish development. The principles of leader development describe goals for what leaders need to be developed to do: leading by example, developing subordinates, creating a positive environment for learning, exercising the art and science of mission command, adaptive performance, critical and creative thinking, and knowing subordinates and their families. The core leader competencies and attributes identified in ADRP 6-22 and the Army Leader Development Strategy (ALDS) provide additional detail of what leaders need to be able to do.

1-6. Supportive relationships and a culture of learning recognize that for development to occur a willingness to engage with others must exist. This tenet relates to two of the principles of leader development: creating a learning environment and knowing subordinates and their families (see ADRP 7-0). Leaders, organizations, and the entire Army must set the conditions for development to occur. Leader development is a mindset incorporated into all organizational requirements and mission accomplishment. Leaders must balance leader development against organizational requirements and mission performance. In operational units and other organizations, development can occur concurrently with training and mission performance, especially when leaders create an environment that places real value and accountability on leader development activities and the Soldiers and civilians to be developed.

1-7. Development occurs through both formal systems and informal practices. Reception and integration, newcomer training, developmental tasks and assignments, individual and collective training, educational events, transition or succession planning, and broadening are all activities where development occurs and should be encouraged. Development involves experiential learning that is consistent with the principle of train as you fight. The performance of duties is always an opportunity for learning while doing. Any experience that shapes and improves performance enhances development.

1-8. Feedback is necessary to guide and gauge development. Formal and informal feedback based on observation and assessment provide information to confirm or increase self-awareness about developmental progress. The Army established performance monitoring, evaluation reports, coaching, mentoring, and growth counseling processes to engage leaders and individuals. Each is essential for development.

THE CHALLENGE FOR LEADER DEVELOPMENT

1-9. The Army must develop leaders comfortable making decisions with available information and prepared to underwrite the honest mistakes subordinates make when learning. These same leaders must also be capable of developing others to be adaptive, creative, professional, and disciplined to execute any mission. Leaders should place emphasis on holistic programs that range across grades from enlisted through senior officers and the Army Civilian Corps.

1-10. Developing leaders involves a holistic, comprehensive, and purposeful group of activities. More than any set of activities, success stems from a culture where leaders with a mindset and passion for developing others use daily opportunities to learn and teach. Leader development occurs at home station, in offices, laboratories, depots, maintenance bays, during exercises, and while deployed. Limited day-to-day interaction with their units and subordinates challenges Reserve Component leaders. At the same time, they benefit from the civilian skills of their subordinates. Reserve Component leaders should use the experience and leadership acquired by their Soldiers from their civilian careers and develop strategies that can be executed on-duty and off, keeping in mind the balance that must be achieved between their subordinate's Army duties, civilian position, and family life. For all cohorts, the Army must sustain the continuous development of future leaders.

1-11. Successful leaders recognize that they must continually develop their subordinates by maximizing opportunities in the institutional, operational, and self-development domains. It is critical to the long-term sustainment of the Army. Leaders are responsible for ensuring their organizations develop subordinates, perform missions, apply doctrinally sound principles in training, and exercise stewardship of resources. Along with responsibility comes accountability. Accountability speaks to two levels: leaders held accountable for how well they have developed their subordinates and individuals held accountable for their own professional development.

1-12. The ALDS lays out the Army's vision, mission, and framework for leader development. The strategic vision emphasizes competence, commitment, character, skills, and attributes needed by Army leaders to prevail in unified land operations and leading the Army enterprise. The Army's leader development mission relies on training, education, and experience components to contribute to the development of leaders. The ALDS also identifies the ends, ways and means for the leader development process. Will and time applied to development are the essential means for success, and this is why a professional culture and individual mindsets committed to development are important. The ALDS starts with leaders at all levels understanding their responsibility for developing other leaders and themselves and creating conditions that provide the opportunities for teaching, training, and providing developmental experiences. The ALDS integrates leader development domains with the training, education and experience lines of effort to show how leaders can be prepared through diverse, aligned activities. The desired ends are leaders developing and improving to meet the expectations identified in the Army leadership requirements model.

LEADERSHIP REQUIREMENTS

1-13. An *Army leader*, by virtue of assumed role or assigned responsibility, inspires and influences people to accomplish organizational goals. Army leaders motivate people both inside and outside the chain of command to pursue actions, focus thinking, and shape decisions for the greater good of the organization (ADP 6-22). These occur through *leadership*—the process of influencing people by providing purpose, direction, and motivation to accomplish the mission and improve the organization (ADP 6-22). The nation and the Army has articulated the expectations of leaders in the Army. The Army leadership requirements model (see figure 1-1 on page 1-4) illustrates expectations of every leader, whether military or civilian, officer or enlisted, active or reserve. This model aligns the desired outcome of leader development activities and personnel practices to a common set of characteristics valued throughout the Army. It covers the core requirements and expectations of leaders at all levels of leadership. Attributes are the desired internal characteristics of a leader—what the Army wants leaders to be and know. Competencies are skills and learnable behaviors the Army expects leaders to acquire, demonstrate, and continue to enhance—what the Army wants leaders to do.

1-14. The competency of getting results requires special mention to counter beliefs that only the end result matters. While the other elements in the model address enablers, conditions, and processes, the achieves category is where leadership is most direct and most challenging. The actions for gets results integrate all other components in a way that brings people, values, purpose, motivation, processes, and task demands together to make the difference in outcomes related to the mission. The integrating actions of this competency also affect all other attributes and competencies. Getting results must simultaneously address improvements to the organization, Soldier and civilian well-being and motivation, adjustments due to situational changes, ethical mission accomplishment, and so on. All the competencies and attributes together lead to trust between the leader and the led, trust that lays the foundation for mission command and effective teamwork.

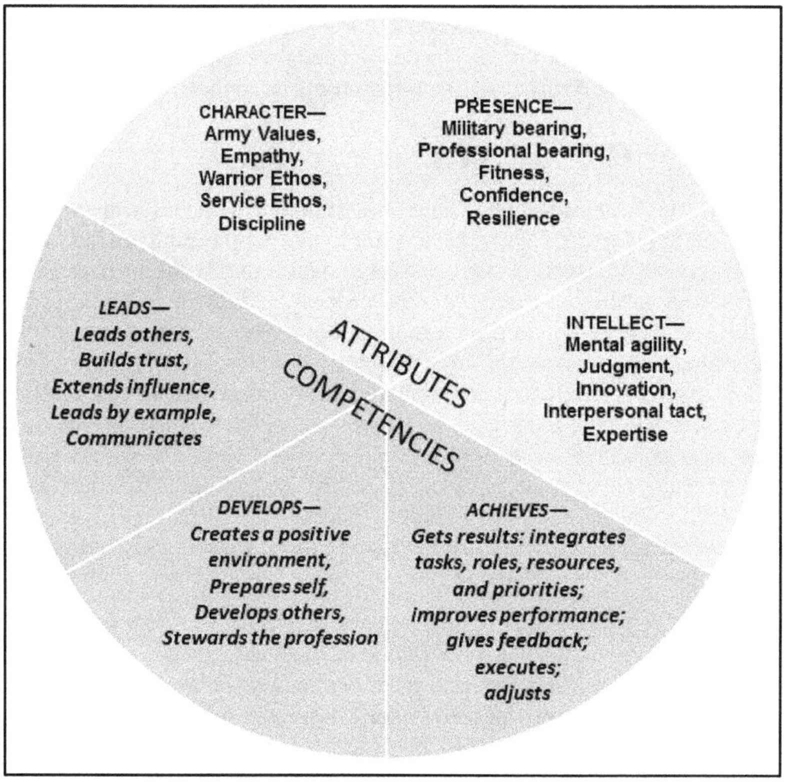

Figure 1-1. Army leadership requirements model

1-15. The leadership requirements and principles of mission command are mutually supportive. Understanding and practicing the principles of mission command are imperative for all leaders: officers, warrant officers, noncommissioned officers (NCOs), and Army Civilians. *Mission command* is the exercise of authority and direction by the commander using mission orders to enable disciplined initiative within the commander's intent to empower agile and adaptive leaders in the conduct of unified land operations (ADP 6-0). While commanders exercise mission command, the actions of subordinates influence effectiveness.

1-16. Through practices in all domains of leader development, the philosophy of mission command becomes ingrained in the Army's ethos and culture. Army leaders, Soldiers, and Civilians at every echelon throughout the operating force and the institutional Army apply mission command principles in the conduct of routine functions and daily activities.

1-17. To best prepare leaders for the uncertainty associated with Army operations, leaders must develop and create opportunities to understand and become proficient in employing the mission command principles. This development requires continual assessment and refinement throughout the individual's service. Leaders who fail to assess or develop their people or teams will not have prepared them to take disciplined initiative. Additionally, the leaders will not understand what individuals and teams are capable of doing and will not be in a position to capitalize on using mission orders.

1-18. Army leaders exercise mission command. Table 1-1 shows the linkage between the principles of mission command and the competencies and attributes of Army leaders in the leadership requirements model. Leader development activities must maintain the vision of developing leaders to execute mission command.

Table 1-1. Principles of mission command linkage to Army leadership requirements

Principles of Mission Command	*Army Leadership Requirements (ADRP 6-22)*
Build cohesive teams through mutual trust	Develops others—builds effective teams. Builds trust—sets personal example; sustains a climate of trust. Demonstrates the Army Values and decisions consistent with the Army Ethic. Leads others—balances subordinate needs with mission requirements. Extends influence beyond the chain of command—builds consensus and resolves conflict. Creates a positive environment—fosters teamwork.
Create shared understanding	Communicates—creates shared understanding. Demonstrates interpersonal tact—interaction with others. Leads others—provides purpose, motivation, and inspiration. Extends influence beyond the chain of command—uses understanding in diplomacy, negotiation, consensus building. Builds trust—uses appropriate methods of influence to energize others. Creates positive environment--supports learning. Gets results—designates, clarifies and deconflicts roles.
Provide a clear commander's intent	Leads others—provides purpose. Communicates—employs engaging communication techniques. Gets results—prioritizes taskings.
Exercise disciplined initiative	Leads others—influence others to take initiative. Demonstrates the Army Values—duty. Demonstrates self-discipline—maintains professional bearing and conduct. Demonstrates mental agility—anticipates uncertain or changing conditions. Gets results—accounts for commitment to task.
Use mission orders	Leads others—provides purpose without excessive, detailed direction. Develops others—expands knowledge. Gets results—executes plans to accomplish the mission the right way.
Accept prudent risk	Leads others—assesses and manages risk. Gets results—identifies, allocates, and manages resources. Stewardship—makes good decisions about resources.

COHESIVE AND EFFECTIVE TEAMS

1-19. Teams are an essential configuration of how people come together to accomplish missions. In the Army, teams occur throughout every structure level of the organization. The Army as a whole is teams of teams. It begins with buddy teams—two military members who look after each other in a variety of positions and environments. The missions of the Army demand that leaders and teams be developed and ready. It is proven that a team is more effective than an individual when members work together, using their unique skills, experiences, and capabilities. The Army leadership competency categories cover how Army leaders lead; develop themselves, their subordinates, and organizations; and bring efforts together to achieve results. Army leaders are charged with developing others and conducting team building. Holistic leader development programs contribute to unit cohesion, resilience, and agility by producing leaders and teams that are creative, life-long learners, adaptable, and capable of exercising mission command.

1-20. The mission command philosophy helps to set the conditions for developing teams. Creating a shared understanding is the first step and most important in developing a team. It gives the team a unifying purpose. The leader sets the tone; in a team-focused climate, members understand how they contribute to the overall success of the organization. Knowing the 'why' drives each action taken. Developing an overall sense of team and building an effective high quality team are two separate actions that should be parts of the overall leader development program. The goal of team building is to improve the quality of the team and how it works together to accomplish the mission.

1-21. The Army relies on effective teams to perform tasks, achieve objectives, and accomplish missions. Building and maintaining teams that operate effectively is essential to both internal and external organizations. To do this, Army leaders employ ***Army team building, a continuous process of enabling a group of people to reach their goals and improve their effectiveness through leadership and various exercises, activities, and techniques***. Figure 1-2 outlines the Army team building process.

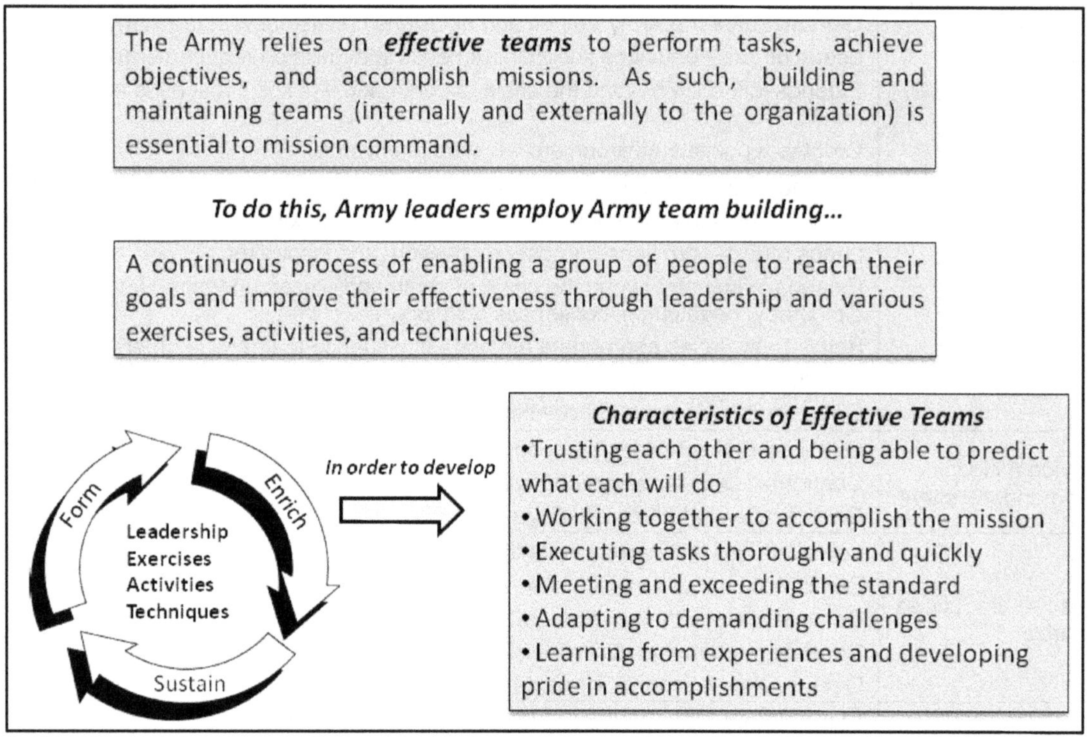

Figure 1-2. Army team building process model

1-22. Three qualities measure good teamwork: identity, cohesion, and climate. Team identity develops through a shared understanding of what the team exists to do and what the team values. Cohesion is the unity or togetherness across team members and forms from mutual trust, cooperation, and confidence. Teamwork increases when teams operate in a positive, engaging, and emotionally safe environment. An engaging environment is one where team members desire to work together on required missions; they feel a sense of self-worth and they are accomplishing something more important than they are. A safe environment occurs when team members feel they can be open and are not threatened by unwarranted criticism.

> **Team building**
>
> From a first sergeant: Team building is a vital part of the Army because Soldiers need to feel as though they are a part of a team if they are going to be willing to fight and die for a teammate and their country. Soldiers need to learn their position and responsibility within that team.

1-23. Teamwork fosters open communication, improves professional relations, and contributes to unit motivation and building trust. Teamwork pulls together the knowledge and experience of a diverse group of people to accomplish the mission. Knowing the elements of effective teams and developing teamwork helps leaders assemble the team, orient them, create an identity, cultivate trust, engage in solving problems, manage processes, regulate team dynamics, and deliver results to other organizations and stakeholders. High performing teams enforce high standards and hold each other accountable for their actions and their level of performance or output. Motivation and discipline are keys to teams that surpass normal expectations. Table 1-2 compares characteristics of effective teams and ineffective teams. Time management and prioritization of effort are important for self and team discipline. Teams that find themselves continually operating in a crisis management mode need to break out of the cycle with better prioritization, effective delegation and dedicated time for planning. See Army doctrine on team building and the Virtual Improvement Center for specific techniques.

Table 1-2. Signs of ineffective and effective teams

Ineffective Teams	*Effective Teams*
Fail to listen to relevant input of a team member. Speak despairingly about other members. Fail to enforce or encourage discipline in the team. Compete, rather than cooperate, with other team members. Argue with other team members in front of counterparts or other individuals. Fail to act or make decisions on issues that have implications for the team. Focus more on self-interest than the well-being of the team. Give less than full effort because of low morale or lack of confidence in other team members.	Emphasize what is common among members rather than focus on characteristics that could cause subgroups to form. Hold a shared vision about operating as a team. Share information that may be useful to other team members. Ensure team members periodically engage in group activities (such as sports, meals, or other off-duty activities). Act quickly to promote togetherness when schisms in the group appear or morale drops. Show appreciation and concern for team members. Act as a team instead of individuals; take pride in team accomplishments.

1-24. The mental dimension is an often overlooked part of teaming and team building. Shared cognition of teamwork includes learning, situational understanding, and critical thinking; motivation is common to all. The leader has a role in building team capacity in each area.

1-25. Teams that have a positive learning culture are eager to understand new areas and current situations. High performance teams are motivated to be inquisitive, to find better ways of doing their work, to acquire new information, and to create new knowledge. Leaders can establish a culture of learning by making learning part of the team's goals. Leaders can build up beliefs in the power of learning by how they demonstrate the value of learning to them personally and how they make learning interesting. Leaders can trigger learning by calling for reflection on shared events and individual experiences.

1-26. Teams that thoroughly engage in active situation assessment and understanding will have the best information available to work on required tasks. Asking questions and sensemaking are valuable processes for teamwork. Sensemaking is a process of creating meaning of an experience through discussion. Individual experiences and insights have greater value through a collaborative situation assessment. The depth of understanding achieved is pivotal in how the team handles complex problems. Developing accurate assessments and the ability to make precise distinctions comes from teams that are motivated to practice at discussing problems critically.

GROWTH ACROSS LEVELS OF LEADERSHIP AND BY COHORTS

1-27. Leaders develop the confidence, leadership, and the competence needed for more complex and higher-level assignments through education, training, and experience gained throughout a career. The Army balances education, training, and experience to develop leaders at all ranks and in all cohorts (officer, warrant officer, NCO, and Army Civilian). While the core leader competencies and attributes remain the same across levels, fine points in application and of expectations change. See ADRP 6-22 for discussions on leadership at the direct, organizational, and strategic levels.

1-28. The processes and expectations for each cohort are similar, while the outcomes are slightly different. Grounded in the Army Values, the Army expects all cohorts to be resilient, adaptive, and creative throughout careers of service to the Nation.

1-29. The Army develops officers, at all echelons, to understand and practice the mission command philosophy to lead and execute unified land operations. The Army expects officers to integrate leader development practices with collective and individual training to accomplish the Army's missions and develop subordinates for future responsibilities. They routinely operate at direct-level interactions with others and work at the organizational and strategic levels to plan, prepare, execute, and assess leader development policies, systems, and practices. Warrant officers serve at all echelons as the primary integrators and managers of Army systems. They bring an unequalled depth of knowledge, experience, and perspective in their primary areas of expertise. Warrant officers, at all echelons, understand and practice the mission command philosophy to execute unified land operations. See DA PAM 600-3 for descriptions of the full spectrum of developmental opportunities throughout a career.

1-30. NCOs are responsible for setting and maintaining high-quality standards and discipline while conducting daily missions and making intent-driven decisions. NCOs serve as standard-bearers and role models vital to training, educating, and developing subordinates. Through training, coaching, mentoring, counseling, and informal interaction, they guide the development of Soldiers in an everyday basis and play a role in the development of junior officers. NCOs, at all echelons, understand and practice the mission command philosophy to execute unified land operations. NCOs advise officers at all levels and are an important source of knowledge and discipline for all enlisted matters. See DA PAM 600-25 for professional development opportunities.

1-31. Army Civilians provide crucial continuity that complements the roles of Soldiers. Army Civilian leaders require a broad understanding of military, political, and business-related strategies, as well as, high levels of managerial, leadership, and decision-making skills. Army Civilians create and practice leader development for other Army Civilians and support the development of military personnel while serving as supervisors, mentors, and instructors. At all echelons, Army Civilians should understand and exercise the mission command philosophy while providing mission-based capabilities to support Army missions. See DOD Instruction 1430.16 and AR 690-950 for specifics.

TRANSITIONS ACROSS ORGANIZATIONAL LEVELS

1-32. Cultural and individual mindsets that promote continuous learning are the cornerstone for creating and sustaining an agile Army. Through activities in the institutional, operational, and self-development domains, personnel obtain education, training, and experiences in order to grow and be able to succeed at positions of greater responsibility. As Army leaders progress in leadership responsibilities, it is necessary for them to develop new mindsets and to refine how they will lead at the next level.

1-33. Understanding key shifts in requirements across the progression of levels, helps individuals prepare for what may be ahead of them and helps prepare others to acquire capabilities for their next level. For the Army, the refinement of requirements across levels helps with management of talent. The Army provides opportunities for developmental experience before assigning leaders to positions of greater responsibility.

1-34. The timing of development is especially important in the military because personnel join and move through a series of alternating and progressive education, training, and operational experiences. The approach applies to Army Civilians as well; however, Army Civilians understand that federal service does not program advancement opportunities for most positions. Army Civilians move across positions based on the governing regulations and laws relating to applying for and filling vacated or newly created positions. Ideally, the best of the direct-level leaders are developed into organizational level leaders and ultimately into strategic and enterprise level leaders.

1-35. A clear framework of leadership requirements provides leaders the basis to assess their strengths and developmental needs and to determine goals for improvement. Created through lengthy study and practice, the Army's leadership requirements model (see figure 1-1) specifically provides leaders with an enduring set of attributes and competencies expected of them. The model provides a consistent reference point throughout the progression of professional and personal development. Leaders must improve in all the leader competencies, become more knowledgeable about the way the military operates, and understand how to operate in complex geopolitical situations. In addition to the leadership requirements model, leaders must grow in their ability to understand, visualize, describe, direct, lead, and assess under differing conditions that change at each level of leadership. As leaders progress, they will experience greater challenges based on the scope of the situation, the consequences and risks involved, and the time horizon. As the scope increases, the number of people and outside parties involved also increases. The consequences of decisions increase, as do the risks that leaders must address. The length of time that leaders' decisions apply tend to increase at higher levels as well as the time over which leaders can apply influence.

1-36. Transitioning to the next stage in a career can be difficult, regardless of demonstration of performance and potential at prior levels. When moving into new positions with different demands, individuals may not perform at a previous high level. Individuals must have a developmental mindset to improve what is within their capability and be motivated to do their best. The Army endorses a culture where individuals continually strive to learn, broaden personal skills, and improve regardless of whether such efforts lead to promotion.

1-37. For military leaders there are six transition points spanning the full range of organizational levels. The changing requirements across levels are illustrative of the relative amount of emphasis needed on certain skills or attributes. Not all levels and transitions apply to all cohorts, military fields, or functions and there will be positions that do not fit neatly into the model. For Army Civilians, there are similar transition points, each of which requires additional leadership skills at progressive levels of responsibility. Personnel begin by managing themselves. Leading and preparing self is something that remains through the entire process no matter where one enters and exits the leadership continuum. In this role of leading self, the individual is primarily a follower. Self-management and self-preparation are important steps in preparing for the initial leadership responsibilities. Six transitions that apply to Army organizations are—

- Leading at the direct level. Initial-entry Soldiers and civilians transition from a focus on self to providing direct leadership to others. Junior leaders learn how to plan daily tasks and activities, understand organizational constructs, and how to interact with subordinates, peers, and superiors.
- Leading organizations. The second transition occurs when leaders begin to lead at the organizational level. This level begins at company, battery, troop, staff, and similar organization levels for Army Civilians. Direct level leadership still occurs at this level, but the leaders become leaders of leaders and will rarely be performing individual tasks, unless out of emergency or in undermanned organizations. Coaching subordinate, direct-line leaders and setting a positive example as a leader are two characteristics that stand out for managers.
- Leading functions. The third transition is from leading an organization (as a leader of direct-line leaders) to leading functions. This level involves directing functions beyond a single individual's experience path. Operating with other leaders of leaders and adopting a longer-term perspective are key characteristics of this phase. Functional leaders typically include majors, mid-level warrant officers, and mid-level NCOs.
- Leading integration. A fourth transition occurs when leaders assume command and leadership responsibility for battalion and similar sized generating force organizations. These leaders must become more adept at establishing a vision, communicating it, and deciding on goals and mission outcomes. They need to find more time for reflection and analysis and value the importance of making trade-offs between future goals and current needs. Positive attitudes related to trust, accepting advice, and accepting feedback will pay dividends during this phase and into the future.
- Leading large organizations. A fifth transition happens when leaders operate at the brigade-equivalent and higher levels of operational and institutional organizations. These leaders develop strategy for organizational and strategic-level operations. They are operating outside of their experience paths while leading others operating beyond theirs as well. Leaders in this phase will only be successful by valuing the expertise and success of others and operating within the multiple layers of their organization. Humility is a desired characteristic of organizational and strategic leaders who should recognize that others have specialized expertise indispensable to success. A modest view of one's own importance helps underscore an essential ingredient to foster cooperation across organizational boundaries. Even the most humble person needs to guard against an imperceptible inflation of ego when constantly exposed to high levels of attention and opportunities.
- Leading the enterprise. A final step occurs in the transition to serving as an enterprise leader. Enterprise leaders must be long-term, visionary thinkers who spend considerable time interacting with agencies beyond the military. This level of leader must be willing to relinquish control of the pieces of the enterprise to strategic and lower-level leaders.

This page intentionally left blank.

Chapter 2
Program Development

2-1. Leader development occurs for the benefit of both individuals and the organization. The Army is known for its success in developing leaders rapidly. Multiple leader development opportunities occur in organizations, though not always used for their learning value. Without intent, plans, or a program for leader development, organizational emphasis on learning is based on commander interest and unit climate. Leader development programs leverage the opportunities for development to address individual and organizational goals for development.

2-2. Commanders are responsible for training and leader development in their units and for providing a culture in which learning takes place. They must deliberately plan, prepare, execute, and assess training and leader development as part of their overall operations. Commanders and leaders must integrate leader development into their organizational training plans or leader development programs.

2-3. Developing Army leaders at all levels, military and civilian, is the best means to ensure the Army can adapt to the uncertainties the future holds. Individuals who feel that the Army and their leaders are interested in them are motivated to demonstrate greater initiative and to engage fully in leader development. Leader development programs that are individualized and that have a multi-leveled aspect are the most effective. The content of leader development programs need to account for the individual's levels of competence, character, and commitment.

2-4. Organizational leader development plans must nest in purpose and guidance of the higher organization's plan. Plans should be consistent with Army enterprise concepts, strategy, and guidance on leader development. Leader development plans should provide guidance to subordinate units yet allow them freedom to determine practices and schedules most conducive to their missions. Plans up and down an organizational structure need to align to create synergy and unity of effort. A battalion leader development plan or equivalent-sized unit will identify specific processes supporting leader development. Generating force organizations headed by a colonel or similar ranking Army Civilian are a good target for leader development plans that detail specific processes. The battalion plan should anticipate the needs of and execution by its subordinate units.

2-5. Variations in programs will occur across echelons depending on the type and size of the organization. For example, a division has greater latitude in selecting leaders for special assignments than does a battalion due to the wider scope of opportunities and larger number of leaders. A Reserve Component unit has fewer training days to plan and schedule team building events, so there may be a greater role for self-development and mentoring. Detached and dispersed units have fewer organic assets to prepare and conduct special events but may have access to external opportunities, such as a training detachment on a university campus.

2-6. The Army holds commanders accountable for unit leader development by regulation (see AR 350-1). Accountability can be included as part of the organizational inspection program (see AR 1-201). Responsibility for leader development cuts across all leader and staff roles. Some examples of roles and responsibilities for developing leaders are—

- Each leader develops subordinates.
- The senior warrant officer, noncommissioned officer, and civilian leaders take ownership for their cohorts' leader development in the organization.
- Each leader (as well as those who aspire to positions of leadership) takes responsibility for their own leader development.

2-7. The next-higher echelon commander, human resources and operations staff, and senior cohort leaders must clarify leader development roles and responsibilities. These individuals directly and indirectly affect the efficiency and effectiveness of leader development.

DELINEATING RESPONSIBILITIES

Efficient implementation of leader development programs depends on a clear definition and allocation of responsibilities across leaders and staff both in and outside the organization. Develop a matrix to document notes on the roles and responsibilities for developing leaders in the organization.

UNIT LEADER DEVELOPMENT PROGRAMS

2-8. Leader development is a mindset and process, not merely an event, reflected by everything leaders do. An opportunity for leader development exists in every event, class, assignment, duty position, discussion, physical training formation, briefing, and engagement. Leader development is a continuous and purposeful process. It is an ongoing process intended to achieve incremental and progressive results over time. Chapter 3 covers the fundamentals of implementing the process to create a leader development culture and to promote a mindset for leader development.

PLAN CREATION

2-9. Various types and echelons of commands and organizations label their leader development guidance with different descriptions such as strategy, philosophy, policy, memorandum, plan, or standing procedure. The title and format are less important than having a good plan—one that aligns with the tenets of leader development: committed organization; clear purpose; supportive learning culture; enabler of education, training, and experience; and feedback. The plan helps to inspire and guide the organization to engage in leader development. Plans that incorporate leader development into daily operations without creating extra events will be well received and have the greatest chance for effective implementation. The imperative of having a plan is to bring attention to leader development, provide focus and purpose, encourage the mindset, set the conditions, show how development should occur, and coordinate efforts across the organization.

2-10. Developing a leader development plan follows the same steps used in the operations process (see ADP 5-0). Planning involves understanding a situation, envisioning a desired future, and planning effective ways of achieving that future. The plan should allow for disciplined initiative by subordinate units and individual leaders. A leader development plan is specific because the outcomes need to address both organizational and individual goals as well as both short-term and long-term goals. The long-term focus extends beyond a military commander's tour and beyond the military personnel's time in the unit. Most Army Civilian leaders are not reassigned based on time, though leader development plans similar to those in operational units can serve their needs. Once the commander's visualization is described and the plan is developed, it directs preparation and execution of the unit's leader development program. The commander and unit leaders lead the execution of the program and assess its progress. The leader development program will create change in the organization and in individuals—it is a living document. As the program creates change and as leaders develop, the plan can be updated.

Understand

2-11. To aid in understanding, command teams can use formal assessments such as command climate surveys, unit Multi-Source Assessment and Feedback (MSAF) 360 assessments, training center after action review (AAR) take-home packages, and command inspection program results to focus on conditions indicating unit strengths and developmental needs. The command team takes these various sources of information along with their own observations and discussions with subordinates and colleagues to determine an appropriate focus.

ASSESSMENT CONSIDERATIONS

Planning a holistic leader development program starts with an assessment. Leaders gain the information needed to shape and inform an assessment from multiple external and internal sources. These are some sources for leaders to consider when developing an assessment:

External:
Review the Army Leader Development Strategy, Army Campaign Plan, and command guidance.
Meet with personnel who focus on the organization's well-being such as the higher headquarters' chaplain, Staff Judge Advocate, Inspector General, other staff, and support agencies.
Review higher headquarters' leader development guidance, programs, and plans.
Review prior command inspection program results.

Internal:
Mission essential task list assessment.
Exercise or deployment results and after action reviews.
Operational and training exercise performance records.
Upcoming events or training calendars.
Organizational climate surveys.
Multi-Source Assessment and Feedback unit rollup report.
Personnel roster and personnel qualification records.
Personal assessment of subordinates' education and experience.
Social media.
Tour work areas and facilities.
Evaluations and support forms.
Initial counseling feedback.
Individual development plans.

2-12. The leadership team may not always have existing formal assessments to use. Additionally, the unit mission or composition may change so those sources may no longer apply. In these cases, leaders align goals with their observational assessments and any changes to organizational mission and goals.

One source to determine an organizational developmental baseline is to schedule and complete a unit-level MSAF event. The unit rollup report provides information on organizational leadership strengths and developmental needs that can focus planning and identifying developmental priorities. In addition, assessed leaders receive an individual feedback report highlighting personal leadership strengths and developmental needs. Individuals can use this information to develop their individual development plan (IDP). During periodic developmental counseling sessions, leaders can review subordinate IDPs to gain insight on current developmental priorities and possible program improvements.

Visualize

2-13. There are several sources to inform decisions about setting the desired future end states for leader development. For the philosophy aspects, the team can examine the ALDS, Army Campaign Plan, and the intent in higher and sister organization's leader development plans. The most important and enduring outcomes are stated in a statement of vision or intent, depending on the preference of the commander.

2-14. An organizational leader development plan establishes the goals for specific end states. Each leader development plan has four mutually supporting purposes. The leader of each organization has a designated responsibility to 1) accomplish the mission, 2) improve the organization, 3) enable personnel to be prepared to perform their current duties and 4) develop leaders for future responsibilities and other assignments. Different from unit training plans, the leader development plan addresses long-term outcomes for individuals and the organization (see table 2-1).

Table 2-1. Goals and end states of the leader development plan

	Individual	*Organization*
Short-term outcomes	Improve personnel capabilities for unit duties	Accomplish the mission
Long-term outcomes	Increase personnel capabilities beyond current assignment	Improve the organization

2-15. Outcomes should address at least these four areas. The planning and execution of the leader development program is a responsibility of the leaders in the organization and the individual. The vision or intent helps to focus and synchronize the leader development actions across the organization to achieve the greatest effects.

Leaders who recognize and approach leader development as a process are able to balance the long-term needs of the Army, the short-term and career needs of their subordinates, and the immediate needs of their organizations to determine how and when to integrate leader development opportunities in already-busy schedules

---

Plan

2-16. To start a plan, the leadership team goes through a conceptual process to consider how to achieve its desired end state. The end state and enduring purpose help guide the detailed phase of planning that involves the selection of activities to emphasize in the unit's program.

Leaders with a mindset, clear-cut vision, and a passion for developing others, themselves, and teams are the most important elements of a successful leader development program. They capitalize on every opportunity.

2-17. The activities cover both unit and individual development for short-term and long-term development. The following factors provide ways to structure a plan:

- Phases of a leader's cycle within a unit.
  - Reception.
  - Integration.
  - Utilization.
  - Assignment rotation within the unit.
  - Transition.
- Unit cycles.
  - Sustainable readiness model.
  - Deployment schedule.
  - Green-amber-red time management and training cycles.
- Cohort programs.
  - Sergeant's time.
  - Preparation for Soldier and sergeant excellence boards.
  - NCO professional development.
  - Warrant officer professional development.
  - Officer professional development.
  - Command team.
  - Civilian leader development seminars.
  - Combined leader development programs.

- Developmental focus—common core for the team and all leaders.
  - Essential characteristics of the profession (see ADRP 1).
  - Command climate (see AR 600-20).
  - Mission command principles (see ADRP 6-0).
  - Core leadership competencies (see ADRP 6-22).
  - Core leader attributes (see ADRP 6-22).
  - Performance qualities, such as adaptability, resilience, versatility, creativity.
  - Core unit mission and functions.
- Developmental focus—career paths for groups of leaders.
  - Career leadership responsibilities (see DA PAM 600-3, DA PAM 600-25, Army Civilian Training, Education, and Development System (ACTEDS)).
  - Career Management Field.
  - Functional area.
  - Army Civilian Career Programs.

2-18. Successful programs integrate formal, semiformal, and informal practices. Policy or regulation direct formal techniques. Addressed in doctrine, semiformal activities are commonly practiced and may be required, but failure to conduct them does not carry punitive consequences. Informal leader development consists of opportunities with a focus on learning. Table 2-2 lists ways to enable learning. Setting conditions for development, goal setting, assessments, and advice and counsel all contribute to improved learning. Table 2-3 on page 2-6 provides additional opportunities and developmental activities. Both tables separate various techniques into formal, semiformal, and informal categories.

Table 2-2. Enablers for learning

Learning enablers	*Formal*	*Semiformal*	*Informal*
Setting conditions	• Integration and reception counseling. • Initial performance counseling.	• Understand individual differences in strengths, interests, potential, and development methods.	• Getting to know and understand subordinates. • Build rapport to enable supportive development.
Goal setting	• Individual Development Plan.	• 5-year plan.	• Short-term and long-term personal and professional goals. • Stretch goals.
Assessment	• Performance evaluation. • Certifications. • Inspection program. • Command climate. • Commander 360° assessment. • General Officer 360° assessment.	• Organizational certifications. • Unit acculturation program. • Core unit mission and functions review. • Multi-Source Assessment and Feedback-Leader 360° for self-assessment. • Unit 360° assessment.	• Day-to-day observations. • Asking others about a leader. • Sensing sessions.
Advice and guidance	• Performance counseling. • Professional growth counseling.	• Mentoring. • Coaching. • Training center counterpart feedback. • Instructor feedback.	• 5-minute feedback. • Peer discussions. • Indirect questioning (What have you planned or done for your development lately? What have done to help a Soldier today?).

Table 2-3. Developmental activities and opportunities

Developmental opportunities	***Formal***	***Semiformal***	***Informal***
Challenging experiences	• Broadening assignments.	• Unit succession planning/ Talent management: • Stretch assignments. • Developmental assignments. • Rotational assignments.	• Opportunities to operate in unfamiliar situations. • Broadening tasks, casualty assistance, staff duty, food service duty.
Group leader development	• Leader Training Program. • After action reviews.	• Officer professional development. • Noncommissioned officer professional development. • Combined events. • Team building events.	• Professional reading and writing program. • Sharing experiences. • Excellence competitions.
Education	• Professional military education courses. • Functional, branch, career program, or special training.	• Scheduling or supporting leaders to attend institutional education	• Encourage utilization of new skills and knowledge of recent graduates.
Self-development	• Structured self-development.	• Guided self-development.	• Self-assessment. • Reflective journaling. • Personalized self-development. • Study and practice.
Collective training	• Incorporate leader development goals and processes into training objectives.	• Team building exercises.	• Shared stories of development.

2-19. From considering the learning enablers and developmental opportunities, the command team will create a plan for scheduling events. The schedule assists those leading and supporting the execution. The schedule maintains a reasonable number of activities and direction of emphasis to help ensure quality. Some events are required, such as performance evaluations and professional growth counseling, and the plan's emphasis triggers other activities. The plan should encourage a mindset where leaders take the initiative to incorporate development into daily activities. Such activities include raising questions in an AAR about what was learned about leadership or asking leaders what self-development they are doing.

2-20. The leader decides the best method to describe the leader development program, such as annual concept, quarterly concept with specific events, or theme based. The plan needs to be synchronized with the overall unit schedule considering the training calendar, significant higher headquarter events that need to be supported, and other significant events.

Execute

2-21. Once completed, the leaders distribute the plan throughout the unit to direct program preparation and execution. Depending on the echelon, the leaders will review subordinate unit plans for leader development. The leadership team sets, directs, and leads the organizational goals, shaping the conditions for individual development. Individual leader development is based on the interest and the effort of individuals who develop others and themselves. It is up to each individual to learn, grow, and develop as an Army professional. An individual's IDP is a personal version of a unit leader development plan. Ideally, individuals and their raters work together to develop the IDP. Execution of the leader development plan can become a regular reported item in reviews, situation reports, and training briefs.

2-22. Leaders must ensure the plan affects development positively. The plan is a way to emphasize leader development and desired outcomes for individuals and for units. Leaders develop the plan with an intent to seeing it through. Reviewed, assessed, and updated periodically, the plan is a living document. Leaders commit to creating an open learning environment where leader development becomes second nature. This occurs when leaders integrate leader development into daily administrative and training events, as well as deployed operations.

Assess

2-23. The leadership team needs to ensure that individual development stays the main effort and that the focus does not become the plan or running events. The documented plan can be either an enabler or a detractor to successful execution and achievement of the desired outcomes depending on the degree of mission command and disciplined initiative.

2-24. Leaders assess implementation and execution against the established vision and end states. Just as assessments help set goals for the unit leader development plan, assessments focused on implementation and execution provide useful information on how well the end states are being achieved and areas for adjustment. The leadership team can also assess whether the vision and end states were adequate or need improvement (see Section IV in Chapter 3).

2-25. Leaders must conduct required leader development activities such as performance evaluations, professional growth counseling, IDPs, and command climate surveys. Assessment of a leader in developing others can occur through reviewing how leaders used formal, semi-formal, and informal activities in the program. The leadership requirements model establishes the expectations for these functions and performance evaluations have provided the mechanism for checking. The 360° leader assessments provide personal feedback to the leader on what they have done to establish a positive climate, engage in developing others, and steward the profession. The 360° feedback provides an opportunity to leaders to address and improve their approaches before evaluation.

2-26. Leader development is a holistic process that occurs every day aligning training, education, and experience to prepare leaders to improve. Leader development is critical to all cohorts—enlisted, officer, and civilian—the source of the Army's future leaders. The process balances long-term Army needs, short-term and career needs of subordinates, and immediate needs of the organization.

Example Programs

2-27. Figures 2-1 through 2-5 provide examples of leader development guidance and programs for units. Figure 2-1 on page 2-8 is a sample battalion plan template followed by an example using that template (see figure 2-2 starting on page 2-9.). Program guidance may have annexes for special events or specific cohort programs. Figures 2-3 through 2-5 (see pages 2-12 through 2-18) show example development programs for battalion NCOs, platoon sergeants, and lieutenants. These example programs illustrate the necessity of developing leaders through daily events and not relying solely on a singular program for development.

SUBJECT: Unit Leader Development Program

1. **References.** *[as required]*
2. **Purpose.**
   - *[Mission of the unit]*
   - *[Importance of leader development to the mission, the Army, and to individual leaders]*
   - *[Desired end state(s)]*
   - *[General application and constraints of this guidance]*
3. **Principles/Command Philosophy.**
   - *[Identify overarching principles or command philosophy]*
4. **Priorities, Focus Areas/Lines of Effort, Key Tasks**
   - *[Enumerate priorities in 1, 2, 3, …]*
   - *[List focus areas or lines of effort and associated key tasks. Choices from this FM, paragraph 2-17 provide options to structure the focus areas or lines of effort. Key tasks may be nested by lines of effort and identify quarterly topics for emphasis]*
5. **Roles and Responsibilities**
   - *[General]*
   - *[Commander / Supervisor]*
   - *[Staff]*
   - *[Subordinate Units]*
   - *[Individual Soldiers and/or Civilians]*
6. **Standard Practices**
   from Tables 2-2 and 2-3, for example:
   - *[Integration, reception counseling, acculturation]*
   - *[Performance evaluation and counseling]*
   - *[Leader professional growth counseling, individual development plan]*
   - *[Attendance policy for professional military education]*
   - *[Individual training]*
   - *[Leader development emphasis in collective training events, training centers]*
   - *[others as desired]*
7. **Unit Activities**
   Define special themes from Tables 2-2 and 2-3, for example:
   - *[Unit assignment practices used for development]*
   - *[Circulation plan for day-to-day observations]*
   - *[Cohort, grade, career management field training and certification]*
   - *[Professional development sessions, topics, schedule/frequency]*
   - *[Multi-Source Assessment and Feedback, Unit 360]*
   - *[Team building events]*
   - *[Unit policy for self-development]*
   - *[Utilization of skills of newly trained course graduates]*
   - *[Reading and writing programs]*
   - *[others as desired]*
8. **Implementation/Effective Dates**

Figure 2-1. Example unit leader development program outline

SUBJECT: Battalion Leader Development Program

1. **References.**
   a. Army Leader Development Strategy.
   b. FM 6-22, Leader Development.
   c. Memorandum, FORSCOM Leader Development Guidance.
   d. Memorandum, XXV Corps Leader Development Strategy.
   e. Memorandum, 83rd Division Command Training and Leader Development Guidance.
   f. Memorandum, 7th Brigade Leader Development Priorities.

2. **Purpose.**

   To enable 1-234 IN BN to deploy worldwide with qualified leaders, on order, in support of global contingency operations. To help ensure the Total Army can successfully adapt to future challenges by supporting the professional development of leaders to their full potential. The desired end state is for all leaders in this battalion to be capable of executing their leadership responsibilities to accomplish missions while improving the organization and preparing personnel to accept greater responsibility for potential future assignments at transition. The battalion leader development program is applicable to individual leaders, units, and staff.

3. **Command Philosophy.**

   Leader development in our unit will be accomplished by how we chose to perform our tasks and conduct ourselves as leaders and as a team. Leader development is a shared privilege of all leaders and personnel. We will use the challenge in all experiences to develop leaders to improve how they conduct leadership. Leaders will guide subordinates to reflect and learn from their experiences. Subordinates will be monitored for their readiness for new challenges. Excellence will be reflected in our approach to our training and mission tasks according to references a-f and not by the number of events on the training calendar. Leader development will be embedded into our daily operations and supported by special activities. We shall create and sustain a culture and mindset of using opportunities to improve others and ourselves. Each day we should encourage professional development, build up each other, and make corrections swiftly and justly.

4. **Priorities, Lines of Effort, Key Tasks**

   Priorities:
   1. Culture of learning. Acculturate new leaders to our developmental philosophy through unit integration and by embracing opportunities for learning.
   2. Junior focus. Assess junior leaders to identify and build their individual strengths and address developmental needs. Direct assessments and development in the fundamentals of leadership in the context of the mission essential task list. Development is ongoing and continues after initial leader certification.
   3. Cohesive development. Work together as teams of leaders to support each other's units to help ensure team leader development and brigade mission success.

   Lines of Effort and Key Tasks:
   1. Assignments and developmental experiences. Plan a patterned sequence of assignments and adjust for optimal development of individuals. Use stretch assignments for high potentials. Add or lengthen developmental assignments for those needing them. Take advantage of opportunities to provide development for subordinates.
   2. Assessments and feedback. Conduct and support MSAF Leader360 and Unit360 events and encourage use of 360 results in IDP development. Develop circulation plans for commanders and first sergeants to rotate to observe their subordinate leaders. Find leaders doing something well and recognize it.
   3. Coach/counsel/mentor. Establish rapport with subordinates by getting to know them and their interests. Make counseling and coaching an integral part of battalion operations. Conduct counseling to achieve impact on the individual's development. Use informal coaching at the point of opportunity. Reserve time to serve as mentors.

Figure 2-2. Example unit leader development program

4. Study events. Nest battalion reading program into division and brigade reading programs. Coordinate quarterly themes of the program to address topics applicable to the following quarter in the quarterly training brief.

5. Roles and Responsibilities

a. All individuals. Engage in the leader development of all battalion personnel, including self. Assist with integration of new leaders in the unit, ensuring that all leaders are proficient in tasks necessary to lead Soldiers in combat. With support from raters develop an individual development plan (IDP) and use leader development opportunities to advance development.
b. Battalion Commander. Plans and executes the battalion program with command sergeant major, staff, and company commander assistance IAW brigade guidance and priorities. Provides feedback to officers on their leader development progress. Reviews and approves assignment patterns for company and field grade officers.
c. Company Commanders and principal staff officers. Assist in the planning and execution of the program. Assess the program for completion and desired impact. Authorized to expand the program into areas deemed necessary.
d. Command Sergeant Major. Supports the commander in planning and executing the battalion leader development program. Senior advisor and organizer for enlisted leader development, including unit assignment patterns and integration of training and leader development practices. Ensures leader development is treated appropriately among unit NCOs with respect to mission completion and unit improvement.

6. Standard Practices

a. Raters conduct timely integration and reception counseling for 100% of new personnel in order to acculturate them into the unit's leader development philosophy and to certify them in leader responsibilities according to ref. d. Assign peer sponsors for newcomers to give them a peer perspective on the unit leader development philosophy.
b. Raters and senior raters will include a performance objective on their evaluation to complete performance evaluation and counseling requirements to standard.
c. Provide professional growth counseling and ensure each individual has a well-thought out and actionable IDP.
d. Identify windows for attendance for each individual to their next professional military education (PME) opportunity and assess impact on unit assignment practices. Incorporate PME attendance in permanent and temporary loss projections.
e. Have individual training conducted so that it incorporates goals for development of future leader requirements. Assess what the individual training contributed towards the goals.
f. Incorporate leader development actions in each collective training event: establish at least one training goal focused on the development of leadership competencies, set conditions and plan assessments for the training goal to occur in parallel with mission training, and conduct group and individual reviews of learning associated with the goal.

7. Unit Activities

a. Create a plan for leader rotation within the unit. Planning considerations include keeping individuals in positions long enough to ensure their stability promotes high unit performance, and that proven, qualified leaders move on to positions of greater responsibility, while other leaders are allowed to continue to develop in current positions or positions better matched for their development. Chart the timing and sequencing of leaders into and out of leadership positions. Use temporary vacancies as opportunities for less experienced leaders to be challenged or broadened.
b. Raters should make a leader observation plan to circulate among their rated leaders. Determine the best times to observe how they conduct leadership and other key duties according to unit training schedules. Take time to provide on-the-spot feedback and coaching. Record observations for future use.
c. Conduct leader certifications within one month of assignment to the unit. Commanders two echelons up will perform the certifications. Follow-on checks will be executed within the first 90 days of assuming a leadership position.

Figure 2-2. Example unit leader development program (continued)

d. Conduct a leader professional development session or team building event for battalion, company, and platoon levels annually. Subordinate units may conduct additional events.
e. Conduct a MSAF Unit 360 event within 6 months of assumption of command by the battalion commander or in coordination with the Commander's 360 or brigade's unit 360. Refine individual and unit goals based on MSAF results.
f. Four hours of duty will be allowed each individual per month to concentrate on IDP activities requiring self-development.
g. Conduct a professional reading and writing program focused on topics selected for application to the following quarterly training emphasis. Identify the purpose of the reading and stimulate thinking about how the reading applies to improving the unit and to developing individual leaders. Readings or lessons:
    1Q FY – trust, team building, effective chains of command.
    2Q FY – indirect influence, negotiation, leading in counter-insurgency operations.
    3Q FY – managing difficult behavior, shared understanding, integrating leader development into individual-collective training.
    4Q FY – developing others, coaching, constructive feedback, self-awareness.

8. Implementation/Effective Dates

This plan is effective upon date signed until rescinded or replaced. The plan will be reviewed for currency quarterly by the S3. All leaders will participate in the execution of the plan.

Figure 2-2. Example unit leader development program (continued)

MEMORANDUM FOR All Battalion NCOs

SUBJECT: Battalion NCO Leader Development Program

1. Purpose. To develop agile, adaptive leaders.

2. Procedures:

a. Post-level: monthly Right Arm Night events, mixers, and socials.

b. Brigade:

(1) Company Commanders day. All company commanders come together for physical training and then attend a session at the host battalion's conference room.

(2) Monthly brigade NCO professional development sessions. Per quarter, the first two months is for master sergeant and above and the third month is for all.

(3) Best Warrior Competition.

(4) Brigade Command Sergeant Major to First Sergeant Mentorship Program: brigade command sergeant major spends two hours with each first sergeant at a training event (in first 90 days, halfway through their tour, and 90 days before departing).

(5) Brigade commander hosts a monthly battalion command sergeants major luncheon.

(6) Brigade commander mentor sessions with company commanders and executive officers, and select warrant officers will occur at the officer's beginning of tour, middle, and end of tour.

c. Battalion:

(1) Monthly leader development classes for promotable staff sergeants and above taught by senior cadre (battalion commander, command sergeant major, executive officer; company commanders and first sergeants).

(2) Monthly leader practical exercises led by senior cadre. Example topics may include executing a change of command or change of responsibility ceremony, conducting an APFT, or conducting a height and weight screening.

(3) Professional reading and writing program assignments for all promotable staff sergeants and above.

(4) Quarterly records review for all promotable staff sergeants and above. Used as a tool to discuss future assignments and schooling, both professional military education courses and civilian.

(5) Annual competitions:

(a) Iron Spartan. A six-event physical fitness competition.

(b) Academic Spartan. A 100 question closed book, written exam on regulations and policies.

(c) Best Warrior Competition (Soldier, NCO, and officer categories).

(6) Monthly physical training days: officer day and senior NCOs day.

(7) Monthly first sergeant breakfast with battalion commander.

(8) Battalion commander and command sergeant major in-brief all new personnel.

(9) Rater and senior rating counseling as required; incorporate monthly counseling packet checks.

3. Monitor Army Career Tracker to ensure everyone has a mentor.

Figure 2-3. Example battalion NCO development program

MEMORANDUM FOR All Battalion Platoon Sergeants

SUBJECT: Battalion Platoon Sergeant Development Program

1. Purpose: To establish a platoon sergeant professional development program and provide guidance for its execution.

2. Intent: Provide opportunities for the battalion command sergeant major to assess the professional strengths and developmental needs of battalion platoon sergeants and develop leaders through one-on-one coaching and mentoring.

3. Procedures:

a. Duration. Each session will be one and one-half (1.5) to two (2) hours long.

b. Location. The location will be at the discretion of the platoon sergeant and will be provided when making the calendar appointment.

c. Participants. This is one-on-one time for platoon sergeants with the battalion command sergeant major. There will be no other participants.

d. Frequency. The first session will occur in the first 30 to 60 days of a new platoon sergeant assuming responsibility. Platoon sergeants will schedule a subsequent session every five to eight months thereafter. Platoon Sergeants will conduct a final session in the last month before relinquishing responsibility.

e. Session time break down (estimate):

(1) The first 30 minutes to one hour of the session will be a sit-down discussion by the platoon sergeant with the battalion command sergeant major in an area where they will not be disturbed. Platoon sergeants must come prepared to discuss the topics in detail (outlined later) demonstrating knowledge of their unit and understanding of their mission.

(2) During the second hour, the platoon sergeant will take the battalion command sergeant major to the selected location. Training should highlight what is unique and interesting to that company's mission. This is the primary focus of the meeting. The platoon sergeant should be showing the battalion command sergeant major how the platoon accomplishes a training event. It is the platoon sergeant's responsibility to know what training is ongoing and where. This is time for the platoon sergeant to highlight platoon systems, conduct, or techniques and procedures.

f. Constraints. Platoon sergeants have full latitude to present prepared information to support the topics of discussion. The platoon sergeant will not speak from a prepared script, but should be able to speak specifics about Soldiers in training, sick call, or other appointments.

4. Topics of Discussion. Topics are intended to be ambiguous and prompt open-ended questions for discussion. Platoon sergeants will not receive a briefing shell to fill out. Platoon sergeants must be prepared to discuss these topics in any order or method chosen by the battalion command sergeant major:

Figure 2-4. Example battalion platoon sergeant development program

a. Describe your mission and strategies to improve how your platoon accomplishes its mission. What are your strengths, weaknesses, upcoming opportunities, and potential threats as they relate to your unit? Where do you want to take your platoon?

b. Discuss your assessment of the training management processes and training record management for your platoon. How well does your company utilize DTMS and what are your company's challenges with the system?

c. Describe your platoon's personnel situation including staffing, certifications, physical fitness (with height and weight compliance), medical readiness, and profiles.

d. What programs do you have in place for professional development? Does your counseling system foster professional development and tailor it for each individual?

e. How do you identify, track, and care for high-risk personnel in your platoon? Describe your approach to using non-judicial punishment.

f. Discuss the external relationships and stakeholders that your platoon has to accomplish its mission. What is your assessment of the strength and benefits of those relationships? What are you doing to maintain or strengthen them?

g. One of the following systems will be selected to discuss in detail; platoon sergeants should be prepared to discuss all of them:

(1) Platoon trends for chapters. What are the highest percentages of chapters? What can we do about it?

(2) What does your platoon do for hip pocket training during down/slow time? Who conducts it? How is it conducted?

(3) Training and Soldier Facilities: what is the status of training areas or facilities, barracks, or any company-operated facilities?

5. Scheduling Implementation:

a. New platoon sergeants will begin this program within 30-60 days of assuming responsibility. This population is the highest priority for scheduling.

b. Platoon sergeants nearing the end of their tour. Platoon sergeants within the last three months of their tenure will schedule their session a minimum of two weeks before change of responsibility. This population is second in scheduling priority.

c. Platoon sergeants midway through their tenure (plan on 24 months being the average tour of duty) have the lowest priority of scheduling.

Figure 2-4. Example battalion platoon sergeant development program (continued)

MEMORANDUM FOR All Battalion Lieutenants

SUBJECT: Battalion Leader Development Program for Lieutenants

1. Purpose. To outline the procedures for executing an officer development program for lieutenants.

2. Scope. This program primarily focuses on the professional development of lieutenants. Captains and majors will also participate in certain portions of this program.

3. General. The most important training we do in this unit and in the Army is develop leaders. Leader development is the process through which we develop the skills, knowledge, and attitudes needed to lead, train, and employ units and organizations at increasing levels of responsibility. Leader development revolves around three pillars that in execution are intertwined: institutional training, operational assignments, and self-development. These three pillars form the foundation that leaders draw from in training and motivate the quality Soldiers in our unit and Army.

a. Institutional Training. Selected officers will have the opportunity to attend formal Army and installation schools and courses to increase their proficiency in specific areas. Attendance at these schools will always benefit both the unit and the individual officer.

b. Operational assignments.

(1) Minimally, Infantry lieutenants will serve in two operational assignments in the battalion. The first and most important assignment will be as a Rifle Platoon Leader. For this reason, Infantry lieutenants will serve about 12 months on average in a Rifle Platoon. This assignment will normally be followed by assignment as a specialty platoon leader (Scouts, Support, or Battalion Mortars), as a company executive officer or as a battalion staff officer. Twelve months is not hard and fast. More important than time is the quality of that time. Nine months in a training-intensive period is more professionally beneficial than 14 months (with 8 months spent in a support or non-training mode).

(2) Non-Infantry lieutenants will serve primarily in duty positions requiring their branch-unique specialties.

(3) Branch-detailed Infantry lieutenants. Such officers will serve as Rifle Platoon Leaders in their initial assignment. Their tour will generally be slightly less than that of a standard Infantry lieutenant (9-12 months on average).Provided it is quality time, this is enough to imbed requisite leadership skills and at the same time does not block the queue for platoon leader time. Following that initial assignment, branch-detailed officers will most likely fill staff positions if they remain in the battalion. In some cases, such officers may have the opportunity to serve in their future specialty outside the battalion. These assignments will be coordinated with the senior officer or commander of the detailed branch.

(4) Assignment patterns for captains and majors will be coordinated with the brigade commander.

c. Self-development. This is an important and effective pillar in the leader development process. Each officer is directly responsible for his own self-development. Self-development programs consist of professional reading and self-study. This is an informal, but intentional, program. By definition, we will not dictate this portion of the officer development program. It should be covered during rater and rated

Figure 2-5. Example battalion leader development program for lieutenants

officer counseling sessions: identifying known developmental needs or areas of interest, goals, and ways to achieve them.

4. Concept.

a. Formal instruction. Self-development is augmented by formal classes that provide additional information on selected tactical and leader related topics. The training schedule will be reflect these classes. These classes will normally take two forms:

(1) Leader Team Training pertains to all officers covering general, non-tactical, and professional topics.

(2) Nested Leader Training pertains to leaders two levels down from the sponsor (lieutenants are the focus for battalion nested leader training). These cover tactical topics along with conceptual, interpersonal, and technical skills.

b. Task list. To focus efforts for leader development, specific tasks for lieutenants are included at enclosure 1. These tasks are designed to round out an officer's development and facilitate integration into the unit. They cover topics other than those normally associated with accomplishing unit training. Lieutenants will work with their company commander or principal staff supervisor to complete these tasks successfully. As a goal, leaders should complete these tasks within 90 days of assignment.

c. Counseling. Professional, routine, and goal-based counseling is an integral part of the professional development process. Company commanders, principal staff officers, and the battalion commander will execute counseling plans to ensure that individual goals are established and professional assessments are provided. Enclosure 1 tasks should be used to develop assessments and monitor professional development of junior officers.

Performance counseling as outlined in ATP 6-22.1 will occur according to battalion policy. Counseling will occur in the officer's work area, not the battalion commander's office. Formal evaluation counseling will be the exception. Officers should be prepared to discuss performance and future goals and objectives. Officers should also be prepared to discuss their self-development program and unit goals. Company commanders will arrange counseling sessions with the battalion commander through the adjutant based on their training schedule. The counseling rotation schedule follows:

- Staff officers and HHC: January, April, July, October.

- Alpha & Charlie Companies: February, May, August, November.

- Bravo & Delta Companies: March, June, September, December.

d. Professional Reading. Professional reading is a valued part of self-development. There are numerous recommended reading lists available. Additionally, technology (distance learning and other web-based applications) allows the easy production and dissemination of training videos on a variety of military-related topics. Additionally, several binders of instructional materials are available for use and review in the S3 shop.

e. Mentorship. Nothing is more effective for professional development than a senior leader taking personal interest in the development of a subordinate. Effective mentorship requires an interested and receptive senior and an equally interested and receptive subordinate. It cannot be forced or dictated. I

Figure 2-5. Example battalion leader development program for lieutenants (continued)

cannot by virtue of rank or position simply state, "I am your mentor". It is much more complicated than that. Senior officers take an interest in junior officers by imparting the benefit of their experience and knowledge. Junior officers should recognize this as a valuable resource and seek opportunities to learn from more senior and experienced officers.

5. Program Responsibilities.

a. Battalion commander.

(1) Serves as the primary trainer and teacher for lieutenants. Certifies that lieutenants are proficient and can execute required tasks to standard.

(2) Plans and executes the battalion program with staff and company commander assistance.

(3) Provides feedback to officers on their leader development progress.

(4) Manages assignment opportunities for lieutenants.

(5) Assists in development of assignment patterns for company and field grade officers.

b. Company commanders and principal staff officers.

(1) Assistant trainer and teacher for lieutenants. Enable lieutenants in completing tasks to standard.

(2) Provide feedback to junior officers on their leader development progress.

(3) Ensure newly assigned officers are briefed and enrolled in battalion programs.

(4) Authorized to expand the program into areas deemed necessary for advancement.

c. Individual officers.

(1) Participate in Leader Team Training and Nested Leader Training.

(2) Develop, with raters, an individual development plan.

(3) Lieutenants will complete certification tasks specified at enclosure 1. The goal for completion of these tasks is within 90 days of assignment.

6. Implementation. This program is effective upon receipt of this memorandum. Many of the tasks listed at enclosure 1 may have already been completed by more senior lieutenants. In this case, rating officers (commanders or principal staff) are authorized to grandfather the appropriate tasks.

7. Conclusion. Development of leaders is the most important thing we do. Our Soldiers deserve nothing less than fully qualified leaders who understand and enforce high standards of mission accomplishment.

Figure 2-5. Example battalion leader development program for lieutenants (continued)

Enclosure 1–Certification Tasks for Lieutenants

Certification area	Initials	Date
1. Sponsorship, Reception, and Integration:		
a. Conduct in briefs with key unit personnel.		
b. Develop a support form with assigned rater.		
c. Read and discuss division and regimental history.		
d. Review and discuss unit officer and NCO rating schemes.		
e. Become familiar with additional duties.		
2. Readiness:		
a. Understand division, brigade, and battalion readiness policies.		
b. Review Soldier readiness files of assigned Soldiers.		
c. Attend a Family Readiness Group meeting (married or not).		
3. Personnel:		
a. Read all division, brigade, and battalion policy letters.		
b. Receive briefing from unit reenlistment NCO. Know reenlistment eligibility date and status of all section personnel.		
c. Review Unit Commander's Financial Report to verify financial entitlements of all section personnel.		
d. Initiate a recommendation for an award.		
e. Understand the installation Army Substance Abuse Program.		
4. Intelligence and Security.		
Read and understand the brigade, battalion, and company crime prevention policies.		
5. Training:		
a. Read and understand pertinent tactics, operations, and branch-specific doctrine.		
b. Develop a leader's notebook that records training status of section.		
c. Conduct a platoon training meeting to prepare for a company training meeting.		
d. Develop a battle-focused platoon physical training program.		
e. Read, understand, and be able to apply the basics of building a live-fire exercise.		
f. Serve as officer in charge and safety officer for a marksmanship range.		
g. Plan, coordinate, and execute a fire team or squad live fire.		
h. Conduct and brief a risk assessment.		
i. Read and discuss squad and platoon battle drills.		
6. Logistics, Supply, and Maintenance:		
a. Conduct an inventory of all platoon or section equipment.		
b. Serve as a report of survey officer (may not be able to complete within 90 days).		
c. Read and discuss the brigade and battalion maintenance policies.		
d. Conduct operator-level preventive maintenance checks and services on each piece of equipment in your section (with appropriate -10 operators manual).		
e. Read and discuss the unit recovery policy.		
f. Read and discuss the battalion policy and the installation regulation on handling and controlling ammunition.		
g. Understand the property accountability system and how it relates to the classes of supply ordering system and maintenance.		

Figure 2-5. Example battalion leader development program for lieutenants (continued)

EVALUATION OF LEADER DEVELOPMENT PROGRAMS

2-28. Developing a set of formal and informal indicators that accurately assess the health of unit leader development in the organization is essential. Leaders can use these locally developed indicators to develop a leader development scorecard (see figure 2-6). Indicators may be different for different types of units, such as operational vice institutional. Employing a red/amber/green status suggests indicators requiring further investigation, which may or may not relate directly to unit leader development efforts. The purpose is to identify trends over time and not react adversely to a single occurrence of an indicator.

Add locally-developed leader development indicators to the unit training brief for subordinate units to track and report on indicators of leader development like other key unit systems (such as training, maintenance, or budget). Refine the measures to those that accurately indicate the health of leader development.

Unit Indicators	Status (red/amber/green)	Action Needed
All key leader positions are filled	amber	Review succession plan for first sergeants
Unit leader changes have little or no detrimental effect on unit performance	green	Sustain job shadow
Multiple qualified candidates competed for last leadership position vacancy	amber	Talk with sergeant major to increase platoon sergeant candidates
A subordinate leader shared a challenge they are experiencing	red	Share a personal challenge
Leader(s) express interest in joining this unit	green	Follow up with G1
Leader(s) express a desire to stay in the unit	amber	1 - Yes; 1 - no - talk with chain of command
Last leader with option to leave the Army was retained	amber	Interview captains on career intentions
Other units requested a leader from this unit	green	Commander selected as general's aide
Unsolicited Soldier comments about their leaders	amber	Talk with sergeant major about unit leaders
A new idea or innovation was implemented	red	Implement feedback from brown bag lunch session
Initial performance of new leaders is high	green	Sustain role models running certification
Overall unit performance is high; no sub-unit is a consistent low performer	amber	Increase unit visits to HHC
Leaders and their units demonstrate lessons learned; few repeat mistakes	green	Sustain personal AARs
Leaders want to discuss strengths and developmental needs	green	Sustain IDP use with counseling
Surveys indicate good morale and climate	amber	Follow-up on climate surveys

Figure 2-6. Example unit leader development scorecard

This page intentionally left blank.

Chapter 3
Fundamentals of Development

3-1. Every part of the Army is vested in maximizing its human capital to prevent, shape, and win in the land domain. Every individual that makes up this capital is—or can become—a pivotal leader. While the Army employs many strategies in the development of leaders, the most influential of these coincide with the time spent in operational assignments for Soldiers and while at work for Army Civilians. Working in real settings—solving real problems with actual team members—provides the challenges and conditions where leaders can see the significance of and have the opportunity to perform leadership activities. Leaders encourage development and learning in their subordinates in every aspect of daily activities and should seek to learn something new every day. Self-development can occur anywhere, so it is an important aspect of development in organizations. Other settings, such as education, can apply the principles that are effective and efficient for development in units. Units and organizations operate in a more decentralized manner than educational and training centers. Decentralization makes the sharing of effective practices necessary and beneficial. Educational institutions and training centers are organizations that can adopt these same leader development principles for their own staffs, students, and trainees.

3-2. The fundamentals of development simplify and span the formal leader development activities that the Army has advocated, such as assessing, counseling, coaching, mentoring, broadening, and team building. The fundamentals are common across formal and informal leader development activities and serve to reinforce an Army developmental culture and a developmental mindset. Other sources provide guidance on techniques associated with the formal activities, such as AR 350-1 on MSAF assessments, AR 623-3 on the evaluation process, ATP 6-22.1 on the counseling process, ADRP 6-22 and MSAF resources on coaching, AR 621-7 and DA PAM 600-3 on broadening, and emerging doctrine on team building.

3-3. Efforts to implement leader development will yield better results if the focus is on methods that have already proven effective. Army leadership requires the establishment of interpersonal relationships based on trust and setting the example for everyone—subordinates, peers, and superiors. In leader development surveys, leaders ranked leading a unit, personal examples, and mentoring as the three most effective ways to develop their leadership qualities. Integrating the fundamentals of leader development into the organization creates a positive, learning climate and builds a mindset among leaders that development is a priority. Experience is a powerful learning tool, however, learning from experience is not guaranteed. As the tenets of leader development convey, learning requires commitment and purpose. For learning to occur, experiences need to be interpreted. This chapter covers setting the conditions for development, gathering and providing feedback, enhancing learning, and creating opportunities where experiential learning thrives.

3-4. The following sections focus on the fundamentals of leader development (see figure 3-1 on page 3-2):

- Setting conditions for leader development. Leaders personally model behaviors that encourage leader development, create an environment that encourages on-duty learning, apply principles that accelerate learning, and get to know the leaders in the organization.
- Providing feedback on a leader's actions. Provide opportunities for observation, assessment, and feedback. Immediate, short bursts of feedback on actual leader actions enhance leader development in operational assignments as well as regular counseling.
- Enhancing learning. Use leaders as role models in the organization. Encourage mentoring, training, reflection, and study. Learning from other leaders is one of the most effective and efficient methods of development.
- Creating opportunities. Modify position assignments to challenge leaders. Be deliberate about the selection and succession of leaders. Integrate leader development across day-to-day activities. Evaluate its effectiveness.

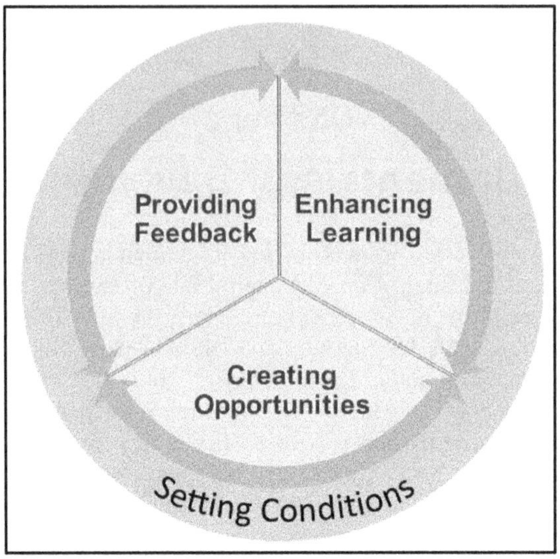

Figure 3-1. Fundamentals of developing leaders

Shared Commitment to Development

From a battalion commander:

Coming into command, I told myself I was going to do it right. I was going to spend seventy-five percent of my time on training and leader development and twenty-five percent on admin. Once in command, no matter how hard I tried, seventy-five percent was admin. To make the change, I had to spread my intent for informal leader development throughout the formation so all understood and were practicing it. Tell those going into command that they have to be deliberate about having a plan for leader development. Then from day one in command, they have to integrate and implement day-to-day, informal leader development or else it will fall by the wayside.

SECTION I – SETTING CONDITIONS

3-5. Establishing a culture that promotes leader development throughout the organization is necessary. The organizational culture needs to embrace leader development to reinforce it as an expected part of daily operations. The culture is affected by leaders who share a mindset that leader development is important and is achievable as part of what they do each day. Leaders need to designate and protect time for leader development and develop a culture that encourages and rewards professional development.

3-6. Leaders create a pro-development culture through the same behaviors they use for any task. Individuals will accomplish what leaders view as important. How leaders react under pressure or during organizational crises will shape what the organization views as important. Leaders' assignment and prioritization of resources also speaks to what is valued and important. What coaching, teaching, mentoring, and counseling that leaders do sends a message of what is important. Leaders' treatment of others through rewards, recognition, and feedback is another sign of what is important. Leaders who demonstrate behaviors supportive of learning and development create a learning environment for the whole organization. ADRP 6-22 describes the importance of leaders setting the example.

3-7. Two other keys are addressed as part of setting conditions:

- Leaders establish a learning environment by encouraging subordinates to take reasonable risks, grow, and develop on their own initiative.
- Leaders gain knowledge of subordinates in the organization as individuals with unique skills, abilities, backgrounds, and goals.

LEARNING ENVIRONMENT

3-8. Leaders set the conditions for leader development by performing their tasks and missions in ways that signal to subordinates throughout the organization that leader development is important. It can have a big effect in return for minimal personal time and resource investment.

Be receptive to individuals input, recommendations, and advice. Be receptive and shut down others who belittle someone's suggestion to take a different or creative approach. Good leaders back subordinates trying to do the right thing and learn something new at the same time. A positive leadership climate encourages a learning environment.

3-9. Being a role model—setting the example—for leader development is essential. Leaders who model these leader actions encourage effective development in others and signal that leader development is valued:

- *Encourage development.*
  - Are you actively and directly engaged in the development of others.
  - Are leaders visibly present and actively engaged in the development of others?
  - Are leaders inspiring others through genuine concern for their growth?
  - Are leaders readily available to provide guidance and answer questions?
  - Do leaders defer to others to develop their subordinates?
- *Encourage learning.*
  - Do leaders feel free to ask themselves what went right and wrong in planning and executing an operation? Is there tolerance for discussing mistakes?
  - Do others observe you engaged in learning? Do you develop yourself? Are you prepared to meet mission challenges as they arise?
  - Do you actively listen to what others have to say? Are followers encouraged to provide candid feedback?
  - Do you create a positive environment? Do you enjoy being a leader and does your example motivate subordinates to emulate you?
  - Do you develop subordinates? Do you know their strengths, developmental needs, goals, and life activities that extend beyond the workday? Are you serving as a teacher and mentor?
- *Promote learning from mistakes.*
  - When mistakes occur, is the focus on assigning blame or on why the mistakes occurred and how to reduce the likelihood of a reoccurrence? Do you avoid criticizing individuals publicly?
  - Do you speak openly about personal leadership mistakes and lessons learned?
- *Encourage innovation.*
  - Are leaders restricted to operating strictly according to standing operating procedures? Do leaders dismiss new ideas in favor of tried and true practices?
  - Do you promote innovation? Can leaders debate with you, exchange issues, or challenge each other's perspectives?
- *Allow for risk taking and encourage exercise of disciplined initiative.*
  - Do you delineate the boundaries or prioritize the areas where subordinates can take risk? Is it clear what is or is not acceptable?
  - Are leaders willing to accept the challenges in unit performance that come with new ideas?
  - Do you show empathy? Do you consider the situations of others relating to their challenges?
- *Encourage effective decisionmaking.*
  - Are leaders well informed when they make important decisions? Do they consider and understand the relevant consequences for Soldiers, Army Civilians, and the mission?

Be aware of personal reactions during times of crisis. Easily remembered, leader behavior sends a strong message about beliefs and values. Before making off-the-cuff remarks, leaders should collect their thoughts and ask themselves what they want less-experienced leaders to learn from this reaction to a crisis. What should subordinates do when they face a similar situation?

Approachability Enables Development

From a battalion staff officer:

An open door policy is critical for demonstrating approachability to subordinates. However, it is not enough just to say you have an open door policy. I think my commander is nonjudgmental and seems levelheaded on any issue, which makes it easy to bring issues to his attention. He is humble, genuine, and patient. At unit functions, I watch him and his wife circulate. He drops down to eye level with the kids and gets a laugh from them for some remark. They both give their full time and attention to whomever they are talking to.

Because of his style, laid-back but concerned and competent, he elicits open communication from subordinates. He starts out by asking about the individual and their family before getting to unit issues by asking, "How's it going?" I noticed that people open up when he uses this approach. He puts them at ease and they know he is genuinely concerned. His approach had a positive effect throughout the battalion.

TECHNIQUES FOR CREATING CONDITIONS CONDUCIVE TO DEVELOPMENT

3-10. Subtle actions on the leader's part build trust and communicate the role of trainer and developer. Experienced leaders use the following techniques to create a developmental culture:

- Be present to observe enough key activities without smothering the leader. After initial observations, give them time and space to exercise leadership without being under the observation spotlight. This helps establish the leader's role as a supportive resource rather than an evaluative note taker. It also builds trust, self-confidence, and creativity in the follower.
- Take an indirect approach. Start by providing descriptions of observations along with positive and negative outcomes. Allow the subordinate to understand what is going well and what needs improvement. The opposite of an indirect approach is to be micromanaging and overly prescriptive, outlining specifics for completion.
- Give each leader a fresh, objective start. Make comparisons between subordinates and an objective standard. Avoid subjective comparisons to past leaders or units (including personal experiences). It is appropriate to reflect on and use personal experience. The bottom line is to observe and assess each leader on individual merit. Avoid thinking of the observation process to grade leaders.

Mistakes occur in all organizations and operational environments. Leaders foster a learning environment by acknowledging that honest mistakes come with challenging missions. Tell leaders about a time you took on a risky, challenging mission. Recount the mistakes made in trying to accomplish it. Wrap up the discussion with the lessons learned from the experience.

> **Taking an Indirect Approach**
>
> From a battalion commander:
>
> I do not walk around the unit with my list of goals to check them off, but I informally encourage leaders to strive for and achieve goals, for both themselves and the unit. For example, during a conversation with a subordinate, I will ask a question like, "What have you done lately to improve yourself?" That catches some by surprise, but it lets everyone know I am serious about leaders setting personal goals and taking action on them. I also ask questions like, "What have you done for a Soldier today, or what have you done to improve the unit?" It is just one question, but it sets an achievement expectation. Sometimes leaders get into a rut where they are executing for the sake of executing; I want leaders to avoid going through the motions since this kills morale.

LEARNING PRINCIPLES

3-11. Development is a process of change. Developmental growth is the same as learning. Learning is gaining knowledge or skill through study, practice, experience, or instruction. Knowing ways to promote learning is key to those who set up and conduct leader development. Applying learning principles throughout leader development practices will accelerate and improve learning. Table 3-1 presents common principles used to design instruction to promote effective, efficient, and appealing learning.

Table 3-1. Learning principles

Principles	*How each principle works to encourage development*
Being task- or problem-centered	Learners are engaged in solving real-world problems. Intellect is stimulated with learning that will affect leader and unit performance.
Activation	Existing knowledge is activated as a foundation for new knowledge.
Demonstration	New knowledge is demonstrated to the learner.
Application	The learner applies new knowledge. Repetition and practice across varying conditions enhances application—through interaction with role models and mentors, from feedback and reflection, and by studying other leaders.
Integration	New knowledge is integrated into the learner's world.

3-12. These principles are important for Army leader development because they are compatible and supportive of learning that occurs while completing duties or during professional development sessions or other modes of learning. Opportunities that challenge the individual and allow learning to occur enhance development in operational assignments, as well as in generating force assignments. Learning best occurs when the area to be learned has real-world relevance; what an individual learner already knows related to the subject is activated; new knowledge and skills are demonstrated to the learner; the skills are actually tried and applied by the learner; and the learner has the opportunity to integrate, absorb, or synthesize new insights or create their own take on the knowledge. Training and developmental projects enhance learning when the learner has an interest in the material and sees its relevance. Learning can accelerate when existing, related knowledge that an individual has comes to mind. Providing an example and using new knowledge enhance learning. The mind absorbs knowledge better when there is time for integration by the learner.

> Survey subordinates on the top three skills they need to improve unit performance or review their IDPs to determine what they need to learn. In doing so, subordinates are motivated and increase their reception to the leader skills they need to learn.

3-13. Purposeful learning starts when learners are challenged to know more and do better. Purposeful learning occurs when practice at mastery of tasks and skills are integrated into leaders' day-to-day activities. Applying the learning principles will result in leaders who actively engage in learning, quickly retain and recall information, and transfer learning to novel situations.

KNOWLEDGE OF SUBORDINATES

3-14. For effective leader development, individual relationships with each subordinate are necessary. Leaders who interact with subordinates on-and off-duty better understand their subordinates' backgrounds and experiences. This may enable discovery of special skills and experiences to support specific mission requirements. Likewise, leaders must avoid establishing or being perceived as having improper relationships. Generally, getting to know subordinates communicates a genuine interest in them as individuals. This builds confidence and generates trust. Trust is key to having candid talks with leaders about their development needs.

There are boundaries to what leaders should know about the personal lives of organization members. Some personal issues may be sensitive and leaders must be aware and understand the sensitivity. Interacting with subordinates in varied on- and off-duty situations enables leaders to build appropriate relationships and develop the trust necessary to discuss sensitive situations.

TEAM TRUST AND UNIT COHESION

3-15. An initial and ongoing objective of a leader is to create a culture that supports leader development. A key accomplishment is for subordinates to accept you as part of the team. This means they trust you as an advisor and coach who facilitates their success. Starting with the first encounter, leaders position themselves as trusted advisors by communicating and modeling attributes and competencies to create a developmental culture. Initial communications might start like this—

- "The only thing I want out of this is to help you (or your staff or unit) maximize capability."
- "I am a developmental resource. The measuring stick for success here is for you to look back when it's all over and see the progress you have made"
- "Tell me a little about yourself—what have you been going through leading up to this assignment? How much experience do you have in your current leadership role? What comes next for you?

3-16. The objective of engaging in this communication with subordinates is as much about listening to their response and understanding their situation as it is about clarifying your role and willingness to be a developmental resource. It is important to build rapport by sharing something about yourself.

From a master sergeant: Without trust, Soldiers will not level with you—at best, you learn either non-truths or part truths. The best way to start building trust is to take the time and talk to your Soldiers from the first day that you become their leader.

3-17. Early in interactions with subordinates, briefly share personal experiences—including areas of specialized expertise and areas of less experience. Candor helps build credibility while at the same time role modeling that it is okay to bring up personal leader developmental needs. It is important to establish trust and a developmental culture. Subordinates have to be receptive, engaged, and ready to develop. With some individuals, it may take extra interaction time to build the necessary level of rapport. Some individuals will seek additional attention and feedback and some will want less.

INDIVIDUAL DEVELOPMENT PLAN (IDP)

3-18. Counseling and feedback provide clear, timely, and accurate information concerning individual performance compared to established criteria. As a part of professional growth counseling and feedback sessions, leaders should help subordinates in identifying strengths and developmental needs. As part of this process, leaders should help subordinates design an IDP. IDPs enable developing an objective approach to professional development. Army Career Tracker (ACT) provides the central location to develop and track IDPs over a service career for both military members and Army Civilians. Reserve Component IDPs should include career development plans that relate to the individual's civilian career as well as Army career and focus on balancing Army careers with civilian careers and family life. Figure 3-2 provides an example IDP.

EXAMPLE

IDP TIMEFRAME					
Status	Draft		**Last Updated**		
Start Date	01 Sep 2015		**End Date**	30 June 2025	
Name	Daniel R. Christopher	**SSN**	xxx-xx-xxxx	**Rank**	2LT
Duty Position	Rifle platoon leader	**DOR**	24 May 2015	**MOS**	11A
MOS Description	Infantry	**UIC**	WABCAA	**Date Assigned**	14 Nov 2015
ASI	5P Parachutist	**SQI**		**LSI**	

SHORT TERM IDP GOALS					
Goal Description	**Activity Type**	**Range**	**Targeted Completion Date**	**Actual Completion Date**	**Status**
Handgun certification	Personal	Short	30 J[illegible] 2016		Pending
Scout platoon leader	Professional	Short	15 Mar 2017		Pending
Battalion Supply Officer	Professional	Short	15 Mar 2018		Pending
Maneuver Captain's Career Course	Professional	Short	30 July 2019		Pending

LONG TERM IDP GOALS					
Goal Description	**Activity Type**	**Range**	**Targeted Completion Date**	**Actual Completion Date**	**Status**
Hindi-Urdu proficiency	Personal	Long	30 July 2020		Pending
Company command	Professional	Long	30 July 2021		Pending
Complete master's degree	Personal	Long	30 May 2023		Pending
Complete Command & General Staff College	Professional	Long	30 June 2025		Pending

Figure 3-2. Example IDP

INSPIRATION SOURCES

3-19. To maintain the momentum of leader development activities, leaders need to reinforce purpose and provide inspiring examples. The Army promotes three reasons for leader development:

- To sustain and improve the immediate performance of the organization. Better leaders translate into better performing teams and units, and better units accomplish their mission.
- To improve the long- and short-term performance of the Army. Better-prepared leaders will be better equipped to fulfill the Army's leadership needs in the future.
- For the well-being of the individual leader. Leader development will let good leaders know that the Army values them and fulfill their desire to learn and to meet personal goals.

3-20. Personal experiences with leaders and leader development that provide inspiration are—

- An exceptional leader, peer, or subordinate who deliberately puts you in challenging situations to grow and learn.
- A leadership challenge where prior experiences prepared you.
- An exemplary professional role model who inspires and motivates others.
- Leaders who, at their own initiative, took responsibility for their own development.

PERSONAL INSPIRATION

Note sources of personal inspiration for investing in leader development. Use these notes to communicate with others and personally understand the importance of developing leaders when distractions threaten implementation.

> **SHARING EXPERIENCES**
>
> **Learning from the experiences of others can be invaluable. The purpose of this discussion is to give leaders the opportunity to share their experiences in terms of the leader competencies (see ADRP 6-22).**
>
> **Discussion Questions:**
>
> **Choose a competency. Discuss the listed behaviors that support it.**
> **Describe a situation where you or someone you observed demonstrated the competency well.**
> **What actions did they take?**
> **What was the outcome?**
> **Why do you consider this a good demonstration of the competency?**
> **Describe a situation where you or someone you observed did not demonstrate this competency well, but could have.**
> **What actions did they take?**
> **What was the outcome?**
> **What actions would have been more effective?**

SECTION II – PROVIDING FEEDBACK

3-21. Leaders need to learn how to observe subordinates and provide developmental feedback. Using multiple methods of assessment and feedback provides a robust and more accurate picture of the individual and provides better developmental opportunities.

3-22. A leader's ability to provide feedback to subordinates will significantly contribute to their development. It will enhance and accelerate learning from the day-to-day work experience—the most valued and effective environment for leader development. Timely, accurate feedback should translate into better leader performance, which will in turn have an effect on unit performance and mission success. Providing accurate feedback starts with planned observation and accurate observation and assessment.

OBSERVATION PLANNING

3-23. The first step to having a legitimate role in a subordinate's leader development process is to observe them. To use available time productively, plan to—

- See them challenged by a developmental need.
- See them excel by applying a personal strength.
- Observe their actions during critical times of unit performance.
- See them reach their limits of strength and endurance.
- Observe decisionmaking and conduct.
- Observe their effect on subordinate leaders and Soldiers.
- See them relaxed and available for conversation.

Do not draw a lasting impression of a leader from a single observation. It usually takes multiple observations before a pattern of behavior emerges. Take time to gather information from others observing the same leader, as different people focus on different aspects.

ACCURATE OBSERVATIONS AND ASSESSMENTS

3-24. Observing other leaders may seem like a difficult task. However, it is a valuable outcome once a leader is familiar with the methods for making accurate observations and providing feedback.

Often a leader will not directly observe the leadership behavior of a subordinate, but will receive a report on unit performance. Leader assessment in this situation requires the leader to communicate the performance indicator to the subordinate. Then, together, move the discussion to the causes of the unit performance. Ask, "What part did your leadership play in the unit's performance?"

Recording Important Observations

3-32. Important details of a leadership observation may be lost or be inaccurately recorded if not written down soon after they occur. Accurate and complete notes are useful when providing leaders with feedback. As described earlier, using the SOAR format is one way to record observations.

Use words that depict action

3-33. A leader needs to describe what the subordinate is doing when they are in the act of leading. By writing down an observation using action words, the leader can be sure the subordinate will be able to recognize it when communicating it back to them. An observation written down using action words appears like this:

Sergeant Jenkins's voice was barely audible and monotone; Soldiers participating in the mission rehearsal could not hear him.

CPT Rider looked directly into the eyes of each platoon leader as he issued the order.

Link to effects and outcomes

3-34. The immediate effect of a subordinate's leadership may be observed in the verbal and nonverbal reactions of others in direct proximity. Leaders and Soldiers in subordinate echelons will feel the positive or negative consequences of a leader's action. Leadership can affect task or mission accomplishment. Trace mission results and look for leader actions that contribute to success or lack of success. There could be a delay in time between the leader's actions and their consequences. The effect may not be obvious for hours or days. The following is a correct example of an observation that includes an effect:

Observation: Sergeant Jenkins voice was barely audible and monotone, so that Soldiers participating in the mission rehearsal could not hear it.

Effect: One vehicle missed a turn on the convoy route; vehicle drove down a road banned from traffic due to IED's. Vehicle attacked by IED. 2 wounded and 1 destroyed vehicle.

FEEDBACK DELIVERY

3-35. When experienced leaders reflect on their own leader development, they place high importance on day-to-day, two-way communication with their senior leaders because they do the same with their subordinates. Feedback is less effective if a leader waits until there is time for a formal sit-down counseling session to provide feedback. Leaders should provide feedback as soon as possible after observing a particular leader behavior to encourage positive effects.

3-36. Day-to-day informal feedback makes sitting down with subordinates for developmental counseling much easier. This informal feedback develops a shared understanding of the subordinate's strengths and developmental needs. Still, many leaders find it difficult to sit down with a subordinate to engage in developmental counseling. ATP 6-22.1 provides extensive guidelines on the counseling process.

Providing feedback on every observed act, response, or behavior will overwhelm a subordinate. Provide feedback based on established competencies and attributes. Focus feedback on a few key behaviors that, if changed, will contribute the most to improved leader and unit performance. Having a focus for improvement will also motivate the subordinate to implement change.

GIVING FEEDBACK IN 60 SECONDS OR LESS

Day-to-day feedback is important to ensure improved leader and unit performance. The following example can guide feedback delivery.

Situation: Commander walks in on a patrol debriefing that one of his company commanders is conducting. He approaches CPT Philips after the debriefing.

Commander provides brief description of the situation: "CPT Phillips, I was in the back of the room while you debriefed the platoon. Let's talk for a minute."

Commander describes the leader behavior observed: "When I came in, Sergeant Jones was describing the suspects he had detained. You listened intently to his general descriptions and asked some pretty probing questions to get details."

CPT Phillips: "Yes sir, I want patrol leaders to understand how important their gathering information is to developing our intelligence efforts."

Commander: "That's a great technique to ask a few questions to confirm what Sergeant Jones is saying and probe for details. He said the suspects were not local. You noticed that. From the excitement in Sergeant Jones's voice, I think he knows the suspects are up to something, but he wasn't sure just what."

Ask the observed leader for a self-assessment before providing personal views. Do this by first recounting back to the leader the situation and observation (the first two parts of the SOAR format). Then ask the leader to provide an assessment and recommendation. This reinforces three important leader development principles: leader self-assessment and self-awareness, individual leader responsibility for leader development, and leader ownership of the recommendation.

DELIVERY OF OBSERVATIONS FOR EFFECT

3-37. It is important to plan how a leader will deliver observations to a subordinate. The delivery methods that follow, when done correctly, provide a leader with an understanding of the effect behaviors have on consequences, all based on careful and planned observations. The two-way communication techniques used for delivering an observation should motivate subordinates to start acting in ways that improve leader and unit performance.

PREPARATION AND TIMING OF FEEDBACK DURING TRAINING

3-38. Before the start of training, leaders should explain the SOAR format or any feedback tool to the unit and its leaders. Leaders should emphasize the developmental nature of the feedback. Armed with this knowledge, unit leadership should be supportive of efforts to deliver of feedback.

3-39. The timing of a discussion of leadership observations can be critical and a deciding factor between whether perceiving a situation as evaluative or developmental. Ultimately, determining the appropriate time for the delivery of an observation is at the discretion of the leader. Consider whether delivery should occur during the action, at a break in action, or at the end of the day or event completion.

During the Action

3-40. Sometimes, leaders deliver observations as they occur. Part of guided discovery learning relies on "during the action" feedback. This is especially true when pointing out to the leader that actions must occur "in the moment" while they can be observed. However, care must be taken not disrupt the training exercise.

Finding a Break in the Action

3-41. Find the right 'break' in the action to deliver observations. This could be during a lull after a major event has occurred (a major success or a failure).

End-of-Day or at Completion of a Major Event

3-42. Consider waiting until the end of the day, especially if observations are lengthy and require discussion. To enable better collective learning, wait until after conducting the unit or team AAR. Then, deliver observations to the subordinate privately, as a mentoring session away from others. This also allows aligning the delivery of observations of the subordinate's strengths and areas for improvement with those of the unit or team as identified in the AAR, assuming they are compatible.

3-43. If observation delivery is best done at the completion of an event, consider letting the subordinate set the time for the discussion. At a minimum, provide a "heads up" about a situation or circumstance to be discussed. This allows the leader an opportunity to reflect and psychologically prepare to listen and receive. This approach reduces the likelihood the subordinate will be preoccupied, nervous, or defensive. Examples of a leader employing this approach include—

- "I'll be back in about 30 minutes and I'd like to talk about how things went this morning. I'm going to ask about how you led the team through the scenario and some of the approaches you took during the decision-making task." [SOAR, Situation]
- "The simulation you led the staff through this afternoon was successful, though I've noted some areas that you could work on. Is there a time you'd prefer to talk later today so I can share my observations and discuss with you?" [SOAR, Situation]

Observation Delivery

3-44. The following steps are an effective way to deliver an observation. These steps represent an indirect approach to providing leadership observations. Once the SOAR outline is completed, leaders are ready to discuss observations and reinforce and recommend actions. The following steps provide a framework for delivering observations, and flow in a logical sequence.

Confirm the Situation

3-45. Start by orienting the subordinate's attention to the observed situation. State the situation and clarify that the observation is about leadership. Reiterate the information recorded: "I would like to discuss the actions you took in the battlefield simulation you just led with your staff." [SOAR, Situation]

Ask for a Self-Assessment

3-46. Ask the subordinate for a self-assessment of the situation and personal leader actions. Guide questioning to the subordinate's leadership during the given event or situation. The subordinate's response should match the leader's assessment. If it does not, the leader should ask additional specific questions:

- "How effective was the communication between you and the subordinates you were leading? And how could you tell?" [SOAR, Associate and Assess]
- "What factors did you observe that may have contributed to miscommunication or a vague understanding among the troops?" [SOAR, Associate and Assess]

Clarify and Come to an Agreement

3-47. Leaders confirm the subordinate either agrees with the assessment or acknowledges a difference in opinion if the subordinate does not share the assessment. Confirm agreement or acknowledgement before proceeding to the assessment, linkages, and observations:

- "That is what I saw as well"
- "Actually, in my observations I noted that you were directive in your message and didn't ask for questions. Would you agree that this is the approach you took?" [SOAR, Observation]

Add your Observations

3-48. Leaders may include observations that the subordinate is not aware. Leaders build on what the subordinate has already said to increase personal self-awareness. Specific behaviors that had an effect on the consequence or outcome include—

- "Your assessment is correct. When you asked for other viewpoints, a good sharing of information followed." [SOAR, Observation]
- "It was clear some of the staff had differing opinions or other points to add, though the opportunity to share really didn't arise." [SOAR, Observation]

WAYS TO FURTHER ENGAGE LEADERS

3-49. Leaders raise questions that will prompt subordinates to think about how to act or respond in the future. Leaders should ask for recommendations about how the subordinate will take better actions in the future, avoid problems, and take advantage of an opportunity. Here are some possible questions—

- "How will you handle a similar situation next time?" [SOAR, transition to Reinforce and Recommend]
- "What steps can you take to avoid this outcome in the future?" [SOAR, transition to Reinforce and Recommend]

Reinforce—Validate a Strength

3-50. Once the leader and subordinate agree on the behaviors that contributed to a consequence and a recommendation for the future, the leader should provide reinforcement on what the subordinate is doing correctly. Here are some examples—

- "Your influencing strategies are working for you, keep it up." [SOAR, Reinforce and Recommend]
- "Consider closing out staff meetings with opportunities for questions or discussion. Your pre-meeting planning and organizing is effective—you should continue that." [SOAR, Reinforce and Recommend]

Additional Tips for Providing Feedback

3-51. There are several other items to consider when providing feedback:

- Focus on the leader's behavior and actions.
- Identify what the leader has control over to change.
- Use focused questions as a form of feedback to create discovery learning.
- Give the leaders the opportunity to come up with a recommendation to the observation. This promotes their taking ownership and responsibility for it.
- Be familiar with improvement actions described in appendix A and offer appropriate ones. Remind leaders that this source is available to guide development, including improving their understanding of positive and negative behaviors and underlying causes.

Providing Feedback on Developmental Needs

From a battalion sergeant major:
It is tough to address or provide feedback on developmental needs, but you have to have a face-to-face conversation right away when someone is not meeting the standard. If not, they might think they are meeting the standard and that I am okay with substandard performance. In general, I make it about strengths and developmental needs of the organization and not a personal attack. This is a career that they have poured their life into, so I am sensitive about that.

Let me share step-by-step how I give people frank, in-the-moment feedback when they need it. How do you get them to engage in a candid two-way conversation and then actually make the changes needed? I start out positive, then talk about the developmental need, then go with something positive again (I term this the sandwich method). I communicate an understanding of the challenges and talk about what they are doing right as well as the shortcomings. Then I pause and ask them for their assessment. We may go back and forth on the issue, getting to what is really going on. Throughout this conversation, I am citing indicators, what I am using to assess them, what I see as a trend. Sometimes, I relate a story about a similar incident or situation to highlight what they can learn from what I saw. In the end, I say, "You own these problems. You can blame it other people, but there is a way ahead." Then we transition to things they could and should be doing. I ask them how they ought to fix the issue, and sometimes they will come up with a better solution than I could have. At other times, I have to go into a "if you do this and this, you can get back on a path to success." I tolerate a certain amount of venting on their part, but in the end, I emphasize that it is their problem to fix. I tell them, "Let's talk about your way ahead, what right looks like, and let's come to consensus on what you need to do."

Lessons from Delivering Observations

3-52. Leaders should avoid delivering some kinds of feedback to a subordinate. These are especially important to avoid—

- Vague and general ideas: "You are a good leader."
- Using absolutes or generalities, such as always or never: "You never follow-up after meetings."
- Observations applied to general traits or the total person: "Your personality is that of an introvert."
- Untimely feedback that the leader is unable to apply: "Two days ago you gave ambiguous instructions at the mission rehearsal."

3-53. It is also important for leaders to learn from the delivery of their observations and realize it takes practice. It is helpful after an interaction for leaders to reflect on their delivery. Self-reflective questions include—

- Was my subordinate receptive to what we discussed?
- Based on my questions, how easily did they identify the behaviors that needed to change?
- Did my subordinate ask for techniques or ideas on how to change or improve?
- Is there agreement on the next step of development and its timeframe?
- Is there evidence that my subordinate is taking action on the observations?

3-54. After delivering observations, leaders look for the next opportunity to observe the subordinate's leadership. Then, gauge how well the subordinate received the observation, what steps the leader has taken to change behavior, and what effect the change is having on unit outcomes.

SUBORDINATE RECEPTIVENESS TO FEEDBACK

3-55. Trust and a developmental culture are critical to ensuring reception of leader observations. If subordinates perceive a leader to be genuinely interested in helping subordinates, the subordinate will be receptive to observations than if there is doubt or mistrust about motives.

3-56. To gauge receptiveness, leaders must remain attuned to verbal and nonverbal cues. These may occur as verbal disagreement or resistance, or nonverbal gestures such as folded arms, rolling eyes, or a lack of attentiveness. Refocus the subordinate by—

- Reaffirming the intent of your feedback is to maximize the subordinate's capabilities to achieve optimal unit performance.
- Reminding the subordinate that your observations are for development—not evaluation or judgment. You are a developmental resource for the leader and the unit.
- Reiterating what went well and note any incremental progress made thus far.

> **Overcoming Resistance to Feedback, Going the Extra Step**
>
> From a battalion commander:
>
> I had one platoon leader who could not do anything right. Falling out of battalion runs was the least of his problems. Whenever you talked to him, he always had an attitude. He was also the one whose fire point you'd roll up to, only to find him and his platoon lying there with no gear on underneath a shade tree, and think, "Okay, enough of this. What is going on here?"
>
> With this particular officer, I moved him from one battery to another thinking that with another commander he would be okay. He started to show little improvements, but then he would slip. He started having moments of brilliance followed by moments of misery. Regardless, I saw something in him. I said, "Okay, he's beginning to get some momentum. How can I help him keep it up?" About that time, part of the unit was deploying. In one unit, the battery commander, first sergeant, and all the platoon leaders were going, and I needed a rear detachment commander for this unit. I could have given it to a senior staff sergeant, but I gave this young officer an opportunity. I sat him down in my office, and frankly said, "I'm going to give you a chance to hit a homerun here, and the good news is you have to hit a homerun. You cannot fail. If you fail, you'll get a relief for cause report card as a commander and might as well find another job." Maybe there was a little bit of a threat there, but he looked at me and said, "Sir, I'm up for the challenge and I'm not going to disappoint you." For the past three months, he has been fantastic—I do not know where he came from.
>
> My sergeant major and I had our fingers crossed for a month that everything would work out, because if he failed, we failed. We took extra steps and precautions to make sure this young man would be successful. We gave him a very strong NCO to be his rear detachment first sergeant. We spent a little more time with him. We tried to set him up for success, and I think it worked. We created a lot of confidence in this young man who was in the doghouse for a long time. Now he thinks, "Hey, I can do this and do it pretty well."

SECTION III – ENHANCING LEARNING

3-57. Setting conditions and providing feedback and advice are two of the fundamentals of development. Applying practices to enhance learning will make development more effective. Enhancing learning draws on the developmental value from learning opportunities. Learning from experience can be enhanced by facilitating what an experience means. Making sense of an experience requires interpretation of the event to create personal understanding. This process requires observation, feedback, dialogue, and reflection. A leader-subordinate pair, coach, or mentor can use these four steps with a leader, any group, or adapted for an individual learner. This section focuses on how dialogue can bolster the process of reflection and understanding. Chapter 4 addresses how an individual uses this learning process. At the individual level, experiential learning is learning while doing. At the organizational level, experiential learning is improving while doing. Experiential learning is consistent with the principle of train as you fight.

3-58. Practical approaches to enhance learning include leader role models, mentoring, guided discovery learning, and individual and group study. These practices are not events that come up on a schedule. They are powerful ways to integrate and promote learning in the day-to-day operations of the organization.

LEADER ROLE MODELS

3-59. Because leaders vary in their skill and experience level, an effective way to learn is directly from unit role models. Positive role models exhibit leadership behaviors that others should emulate. Leveraging role models for leader development is an efficient use of time and resources. They are a resource right in the organization. Supervisors should identify role models for each key position (such as company commander or platoon sergeant) and may want to identify role models possessing special skills that other leaders need to master. Leaders should resource these role models appropriately for the responsibilities. Likewise they should create opportunities for less experienced individuals for interaction. For example, supervisors may assign—

- A role model to new leaders for their reception and integration.
- A role model to coach a leader due to possessing a particular skill or special expertise.
- Role models to run leader certification programs.
- An inexperienced leader to shadow a role model for a specified period.

THE 5-MINUTE SHADOW

Bring in a subordinate to observe or participate in an aspect of work that will make them a better leader. To maximize the experience—
- Communicate the situation, decision, or issue.
- Convey the importance of acting appropriately or making the right decision.
- Describe possible consequences, second- and third-order effects.
- Discuss the decision or actions and reasoning behind them.

MENTORSHIP

3-60. Mentoring can benefit leader development efforts. *Mentorship* is the voluntary developmental relationship that exists between a person of greater experience and a person of lesser experience that is characterized by mutual trust and respect (AR 600-100). A mentor is a leader who assists personal and professional development by helping a mentee clarify personal, professional, and career goals and develop actions to improve attributes, skills, and competencies. A mentee is the individual receiving mentorship. Individuals are encouraged to participate in mentoring as a voluntary experience. Age or seniority is not a prerequisite for providing mentoring. A junior individual may mentor a senior individual based on experience or specialized expertise as a subject matter expert.

3-61. Mentoring will occur while individuals are in operational and institutional assignments; however, the mentor-mentee connections are best if they occur outside the chain of command. This is not contrary to the requirement that superiors have the responsibility to develop subordinates. It is differentiating between the role of a mentor and the role of a leader to develop, counsel, teach, and instruct subordinates. Supervisors should refrain from appointing mentors or formally matching individuals with mentors. Participant self-selection leads to the most effective mentoring relationship.

3-62. Leaders foster mentorship by—

- Educating leaders in the organization on mentor responsibilities.
- Participating as a mentor.
- Inviting experienced leaders to visit and share their mentoring experiences.

Senior experienced leaders will visit occasionally. Schedule time for them to meet with a less experienced group. Provide some structure to this group mentoring experience by having members generate questions in advance. Their experience and perspectives offer new ideas for focusing development. Communicate intent with the experienced leaders and provide the questions to them.

Mentor Roles and Responsibilities

3-63. Selection as a mentor is a compliment to one's professional abilities and competence. Table 3-2 highlights the general roles and responsibilities of mentors.

Table 3-2. Mentor roles and responsibilities

Role	*Responsibility*
Provides	Encouragement and motivation. Candid feedback about perceived strengths and developmental needs. Advice on dealing with obstacles. Guidance on setting goals and periodically reviews progress.
Shares	Experiences that contributed to personal success. An understanding of the Army, its mission, and formal and informal operating processes.
Encourages	Appropriate training and developmental opportunities. Sense of self-awareness, self-confidence, and adaptability. Efficient and productive performance.
Serves	As a confidant, counselor, guide, and adviser. As an advisor for career development ideas or opportunities. As a resource for enhancing personal and professional attributes.

3-64. Mentoring is a powerful tool for personal and professional development. Mentoring generally improves individual performance, retention, morale, personal and professional development, and career progression. Mentoring offers many opportunities for mentors and mentees to improve their leadership, interpersonal, and technical skills as well as achieve personal and professional objectives.

3-65. It is not required for leaders to have the same occupational or educational background as those they coach or counsel. In comparison, mentors generally specialize in the same area as those they mentor. Mentors have likely experienced what their mentees are experiencing or will experience. Consequently, mentoring relationships tend to be occupation-specific, focused primarily on developing a better prepared leader.

Mentoring Relationships

3-66. The appearance of favoritism or creating conflict with raters or senior raters should keep leaders from mentoring subordinates within their chain of command. Subordinates should avoid approaching superiors in their chain of command to be mentors.

3-67. Mentoring relationships can be described by purpose and relationship:

- Traditional mentoring. Focuses primarily on the mentee, examining the career path through goal setting, with overall development of the individual as the focus. This mentoring is a process where the mentor and mentee join by their own volition.
- Peer mentoring. Occurs when a mentor has extensive knowledge and experience but not higher rank or grade than the mentee. Mentoring relationships may occur between peers and often between senior NCOs and junior officers. This relationship can occur across many grades or ranks.

3-68. Regardless of purpose, a successful mentoring relationship is based on several elements:

- Respect. Established when a mentee recognizes desirable attributes, skills, and competencies that the mentor has and when the mentor appreciates the attitude, effort, and progress of the mentee.
- Trust. Mentors and mentees should work together to build trust through open communication, forecasting how decisions could affect goals, frequent discussion of progress, monitoring changes, and expressing enthusiasm for the relationship.

- Realistic expectations and self-perception. A mentor may refine the mentee's self-perception by discussing social traits, intellectual abilities, talents, and roles. It is important for the mentor to provide honest feedback. A mentor should encourage the mentee to have realistic expectations of their own capabilities, present and potential position opportunities, and the mentor's offerings.
- Time. Set aside specific time to meet; do not change times unless necessary. Meet periodically to control interruptions. Frequently check in with each other via calls or e-mail.

MENTORING BENEFITS

3-69. Soldiers and Army Civilians who seek feedback to focus their development, coupled with dedicated, well-informed mentors, will embed the concepts of life-long learning, self-development, and adaptability into the Army's culture. The benefits are threefold: for the mentor, the mentee, and the organization.

Mentor Benefits

3-70. Serving as a mentor can provide many benefits, such as—

- Professional development. Becoming identified as someone who develops or mentors well-known performers can attract qualified, high-potential individuals who will look for opportunities to work for the mentor. Developing others to follow in a mentor's footsteps can facilitate the mentor's own personal and professional development and career progression.
- Knowledge. Mentees can be a source of general organizational data, feedback, and fresh ideas. Because higher-level positions isolate some executives and managers, mentees can serve as an important link in keeping communication lines open. While the mentor might possess facts about issues, mentees often provide important feedback about views at different levels of the Army.
- Personal satisfaction. Mentors generally report a sense of pride in watching mentees develop and a sense of contribution to the Army. It is an opportunity to pass on a legacy to the next generation.
- Sharpened skills. Mentors sharpen management, leadership, and interpersonal skills as they challenge and coach the mentee.
- Source of recognition. Good mentors are well respected.
- Expanded professional contacts. Mentors develop rewarding professional contacts by interacting with other mentors, supervisors, and contacts made through the mentorship relationship.

Mentee Benefits

3-71. Mentees gain tremendously from a mentoring relationship. Such benefits include—

- Increasing self-awareness through candid feedback.
- Building confidence and encouragement to grow beyond usual expectations.
- Having a role model and a trusted advisor.
- Gaining better understanding of the Army and what is required to succeed and advance.
- Gaining visibility through opportunities to try advanced tasks and demonstrate expanded capabilities.
- Reporting greater career satisfaction with higher performance and productivity ratings.

Organizational Benefits

3-72. The organization and the Army as a whole benefit in the following ways—

- Increased commitment and retention. Mentoring increases the understanding of how to reach the next level of responsibility—enhancing job satisfaction and reducing reasons to leave the organization.
- Improved performance. Both mentors and mentees have an opportunity to expand their technical, interpersonal, and leadership skills through the mentorship relationship. Mentoring helps mentees identify and prepare for positions which best fit their needs and interests. This benefits the Army by enabling it to fill positions with the most capable, motivated personnel. Mentoring is

functionally efficient, because instead of floundering on their own, mentors help mentees to develop career road maps.

- Leader development. Mentoring increases the effectiveness of leader developmental activities that occur within the chain of command and generally produces leaders comfortable with the responsibilities of senior level positions.
- Leadership succession. Mentoring facilitates the smooth transfer of Army Values, culture, traditions, Warrior Ethos, and other key components to the next generation of Army leadership.
- Recruitment. An Army-wide mentoring program makes the Army attractive to potential recruits because it shows the Army cares about its people and their development.

The Mentorship of George C. Marshall

General of the Army George C. Marshall served as the U.S. Army Chief of Staff during World War II. As such, he played a leading role in the expansion of the Army during the war and the selection and advancement of key leaders. Long before assuming his duties as Chief of Staff, Marshall served in important positions during World War I at the division, corps, and American Expeditionary Force staff levels. From these vantage points, he witnessed the difficulties inherent in training large bodies of troops and selecting and developing leaders on a shortened timeline.

In the interwar period, Marshall was determined to avoid the shortcomings that marked the American experience in World War I. He became the Assistant Commandant of the U.S. Army Infantry School at Fort Benning in 1927, a position from which he could instill lessons from World War I. During the next five years, he revamped the school curriculum, changing its focus from following rote formulas to stressing tactical improvisation and creativity based on experience. He chose officers like J. Lawton Collins, Joseph Stillwell, and Omar Bradley to be instructors at Fort Benning, mentoring these men to ensure his approach to training and tactics was passed to younger officers and thus across the Army.

Marshall's approach was important because during his tenure at Fort Benning, over 200 future generals passed through the school either as students, instructors, or both. His approach to training and leadership influenced them all. Marshall later reflected that the officers with whom he served at Fort Benning were the best he worked with during his career.

In addition to his Fort Benning posting, his other interwar assignments, particularly several tours at the War Department and a posting to the Army War College, gave him a broad knowledge of the interwar Army officer corps. This was a key factor in his appointment of senior officers during World War II. Marshall either picked or recommended many of the American generals given top commands during the war, including European theater commander Dwight Eisenhower, China theater commander Joseph Stilwell, Army Ground Forces chief Leslie McNair, army group commanders Mark Clark, Omar Bradley, and Jacob Devers and army commander George Patton. Marshall's mentorship of subordinates at Fort Benning allowed him to select or recommend officers for key positions that he had trained and prepared in practical tactics. This mentoring helped ensure that the Army fighting World War II was able to adjust rapidly to battlefield realities.

Mentoring Responsibilities

3-73. Mentoring is a professional relationship. As such, mentors and mentees each have responsibilities to ensure the relationship is positive and productive. Both must consider their own, and the other person's, interests and expectations. Mentors should follow standards for appropriateness expected of all Army leaders. While personal rapport and candid feedback are both characteristics of good mentoring relationships, mentors should try to focus on professional development advice. A mentee should never use a mentor to bypass normal and appropriate procedures or chains of command or to exert pressure or influence on an individual, such as the mentee's supervisor, who is the appropriate decision authority for the mentee.

3-74. Mentoring is crucial to development and retention. Once the relationship is initiated, the mentor has responsibility to—

- Share organizational insight gained through knowledge and experience. Showing mentees how you have managed a certain situation is far more effective than just talking them through it.
- Expand the mentee's network. Mentors give advice on a spectrum of topics, ranging from specific skills to broader issues of career goals. Mentees gain sound guidance, access to established networks, and enhanced personal and professional perspectives.
- Help with setting development goals. Mentees often seek mentors to enable professional growth, perhaps advancement in the organization or in changing career fields.
- Provide developmental feedback. Giving feedback increases the mentee's self-awareness, particularly concerning strengths and developmental needs. If properly given, feedback results in greater rapport. Positive feedback comes as, "I think you did a good job with the meeting." It tends to be from the perspective of the giver. Turn negative or constructive feedback into "you" statements: "You need to create an agenda for each meeting."

> **Candor in Mentorship**
>
> From a warrant officer:
>
> The best thing my mentor ever did for me was be brutally honest with me. I was a senior chief warrant officer 2 and thought I knew everything. Then there was a situation where my battalion commander and I did not see eye-to-eye. I explained the situation to my mentor, a senior warrant officer. I believed he would side with me and make me feel good. Instead, I got a reality check. He told me that I was the one screwing up. After the shock and being offended that he did not side with me, it jolted me to think. A good mentor has the candor to tell you the bad things you are doing as well as the good, and set you on the right path again. I appreciated this lesson that I still remember, and I use the same approach today with the Soldiers I mentor.

3-75. The mentee must be an active participant in the relationship. In particular, mentees must—

- Prepare. Complete appropriate preparations for meetings with the mentor.
- Develop. Work to achieve the best attributes, skills, and competencies.
- Be flexible. Listen to the mentor and consider all new options proposed.
- Take initiative. Seek the mentor's advice when needed.
- Focus on the goal, not the process. If unclear, ask the mentor how the process leads to the goal.

MENTORING SKILLS

3-76. It is important to possess key mentoring skills to be effective as a mentor. These skills include—

- Listening actively. Focusing on the mentee's main points and whole meaning. Watch body language, maintain eye contact, and understand which topics are difficult for the mentee to discuss. Showing someone that you are listening is a valuable skill. It shows you value what the person is saying and that you will not interrupt them. This requires patience and a willingness to delay judgment.
- Holding back judgments. Reduce emotional reactions (such as anger or excitement) to the mentee's comments. Do not immediately draw conclusions about whether the meaning is good or bad until you are sure you understand the comments.
- Asking the right questions. The best mentors ask questions that make the mentee do the thinking. However, this is not as easy as it sounds. Simply, think of what you want to tell the mentee and frame a question that will help the mentee come to the same conclusion on their own. To do this, try asking open questions that a simple yes or no cannot answer. Alternatively, ask direct questions that offer several answer options. Then ask the mentee why they chose that particular answer.
- Providing feedback. Do this in a way that accurately and objectively summarizes what you have heard, but interprets things in a way that adds value for the mentee. In particular, use feedback to

show that you understand what the mentee's thinking approach has been. This is key to helping the mentee see a situation from another perspective.

- Resisting distractions. Control the location and minimize outside noises or people as much as possible. Focus on the mentee.

GUIDED DISCOVERY LEARNING

3-77. Underpinning all developmental activity is the accurate observation of performance. Armed with accurate observations, the senior leader engages the subordinate in effective two-way communication to deliver observations on actions and behaviors. Effective delivery techniques foster leader acceptance, ownership, and action.

3-78. Besides directly delivering an observation, leaders can use indirect methods. Indirect methods place increased responsibility on the subordinate to identify personal strengths and developmental needs. Indirect methods employ the techniques of guided discovery learning. The techniques are designed to engage subordinates to discover their learning needs, supported by the senior leader.

3-79. Guided discovery learning is an advanced technique that experienced leaders employ to help the subordinate learn. The technique can be used in coaching, counseling, and mentoring situations. Guided discovery learning is effective because—

- It is the subordinate's responsibility to make sense of incoming information and integrate it with their personal base of experience and knowledge of relevant doctrine. It is a process of discovery for the leader.
- Subordinate learning and transfer of knowledge are maximized because the supervisor generally keeps the subordinate on track through hints, direction, coaching, feedback, or modeling.
- Guided learning enables deep understanding of targeted concepts, principles, and techniques.

3-80. Pure discovery learning is less effective than guided discovery learning. With discovery learning alone—

- The subordinate merely executes based on personal experience or knowledge.
- The subordinate makes sense of incoming information using whatever criteria they feel is relevant.
- The supervisory leader is passive, providing no guidance or feedback concerning the rules or criteria that the subordinate is using for problem solving.

3-81. Guided discovery learning is more effective than prescriptive methods where the leader gives the subordinate the correct answer to a problem. Prescriptive methods require neither thinking nor deep learning by the subordinate. They merely execute the prescribed solution given by the supervisory leader.

GUIDED DISCOVERY LEARNING TECHNIQUES

3-82. Guided discovery learning techniques are an effective way to deliver leadership observations. These methods are commonly employed when developing the leadership skills of subordinates:

- Positive reinforcement.
- Open-ended questioning.
- Multiple perspectives.
- Scaling questions.
- Cause and effect analysis.
- Recovery from setbacks.
- Use experience.

Positive Reinforcement

3-83. The first observations of a subordinate ought to focus on what they are doing right. Commenting on positive actions up front shows a commitment to balanced and fair observation. It also builds confidence and confirms performance that is productive and accomplishing an objective.

Open-ended Questioning

3-84. Asking open-ended questions gets subordinates thinking about the situation and their leadership pertaining to unit performance. Broad questions maximize the potential for discovery. Leaders may need to ask additional specific questions if the subordinate is not identifying issues that need attention.

3-85. An advantage of this approach gives subordinates hints about what they may need to do differently, yet allow them to discover the actual issue on their own. In this way, responsibility for evaluation is with the subordinate, as is ownership for fixing the situation.

3-86. Open-ended questioning is useful when the leader has time to listen, reflect, and do something about the situation. Thus, the busiest part of mission planning or execution may not be the most appropriate time to ask an open-ended question—unless it has a critical connection to reflective thought.

> **Coaching through Open-ended Questions**
>
> From an Army Civilian division chief:
> My boss had me write information papers and give presentations on various topics that expanded my knowledge of our organization and the role of my fellow division chiefs. He asked questions that led to shared expectations. He asked who I thought I could seek assistance from to help accomplish those goals. In general, his hands-off approach allowed me to make key decisions for my section and to learn from my mistakes. His approach was always more as a coach than a dictator.

3-87. Open-ended questioning is employed by—

- Identifying the outcome for the leader to realize.
- Asking general questions about factors related to that outcome.
- Asking specific questions and providing hints until the leader connects the outcome with actions.
- Listening closely to the leader's response.
- Confirming and reinforcing what is heard as an accurate assessment.
- Probing further or offer outcome-based evidence if they are not accurately assessing the situation.

Multiple Perspectives

3-88. Employing multiple perspectives helps a leader see the situation they are in from another person's perspective (or a different frame of reference). A complementary step to the decisionmaking process is to understand a problem and appreciate its complexities before seeking to solve it. Supervisors help subordinates reframe the current situation through open-ended questions or soliciting third party feedback from other stakeholders.

3-89. The purpose of multiple perspectives is to prompt subordinates to think creatively and innovatively in their approach. Leaders should use this technique when a subordinate appears to be stuck in a limited way of thinking or is unable to break away from a mental block.

Scaling Questions

3-90. The scaling questions technique is useful in facilitating a leader's self-understanding of how difficult or challenging a problem is in relative terms. It also facilitates incremental improvement and helps an individual recognize that progress has occurred. Supervisors ask subordinates to use a 10-point scale (where 10 is highest or best and 1 is lowest or worst) to assess personal performance on an action or behavior (competencies). The subordinate could share what they could do differently to improve performance one or two points on the scale.

Cause and Effect Analysis

3-91. Leader actions are often several layers or processes removed from their consequences. The cause and effect analysis is a method to identify the root (or original) cause of consequences and outcomes.

3-92. It is not always obvious to leaders how certain behaviors affect outcomes further down the line. Cause and effect analysis is important because a leader and unit will continue to experience a negative outcome until identifying and resolving the actual root cause. Many times only subsequent effects (or symptoms) of a problem are addressed, leaving the root cause intact.

3-93. Leaders use cause and effect analysis when there is limited time and capability to address shortcomings. Identification of a root cause focuses on remedial actions that will fix the problem and change the consequence.

3-94. Cause and effect analysis is facilitated through—

- Asking "What (rather than why) causes it to happen? Show consequences or outcome data.
- Continuing to ask "What?" and "What else?" until identifying all causes. It helps to capture work on paper or a whiteboard.
- Depicting the relationships between causes and effect.
- Identifying which causes, if changed or isolated, would prevent reoccurrence of the outcome or consequence.
- Identifying solutions or changes to implement without causing other negative outcomes or consequences to occur.
- Coaching the leader on being proactive to avoid negative outcomes before they occur.

Recovery from Setbacks

3-95. When a subordinate experiences a difficult situation, setback, or seemingly insurmountable challenge, a supervisory leader can help restore confidence and prevent conditions from going from bad to worse. Employing the following enables recovery from setbacks—

- Reinforcing a strength—a leadership behavior the individual is performing well.
- Helping the leader recognize that they are already successfully handling some part of the task.
- Asking open-ended questions to increase situational awareness and probe for solutions.
- Providing recommendations if or when the leader is unable to arrive at a suitable course of action.
- Increasing the percentage of positive reinforcement and support relative to negative reinforcement.

Use Experience

3-96. By virtue of position and experience, a leader often knows something is going wrong or right before the subordinate knows it. There is an art to knowing when to impart aspects of that experience to a subordinate. A great deal of learning can occur by providing leaders with hints and bits of information, well short of full understanding.

3-97. Leaders should carefully weigh the pros and cons of providing a subordinate with hints during training exercises. It is important to allow situations and events to unfold without premature intervention. If the leader provides information or solutions to the subordinate too soon, the situation's development value diminishes, as situations of ambiguity and adversity compel leaders to adapt and problem-solve on their own.

3-98. Yet, leaders do not want to hold on to information that may inhibit learning during the exercise. Without hints, a subordinate may experience a situation and its consequences, but not effectively learn from it. With hints and additional information, the subordinate launches on a learning expedition while the situation is still evolving. The inquisitive subordinate will follow up on the leader's hints and find out why systems or people did not perform to expectations, a valuable learning expedition.

Special Situation: Working with Non-responsive Indivdiuals

3-99. There will be times when a subordinate does not respond to any of the feedback or discovery learning methods. When this occurs, the leader might reflect on why this is occurring and if there is anything to do differently to trigger desired responses. Ultimately, the responsibility for learning lies with the subordinate. Even in difficult situations, there are techniques to use that may gain the subordinate's attention and create learning opportunities.

3-100. These are some ways to redirect a non-responsive subordinate—

- Redirect efforts to work with the subordinate leader's subordinates and peers—they are likely feeling the consequences of the subordinate leader's behavior. Support the subordinate leader's adaptation to the identified developmental needs by providing solutions and taking action to mitigate effects on the unit's mission. The subordinate leader should notice the change in mission performance and want to know why it is occurring.
- Resources permitting, have the non-responsive subordinate swap places with a peer or have another leader observe the non-responsive subordinate. Compare notes and see if your observations are consistent with that of the other leader.
- Use experience. Talk the situation over with other leaders skilled at observing leadership. Obtain their perspectives and ideas on how to work with non-responsive subordinates.

OBSERVATION, DELIVERY, AND DISCOVERY LEARNING INTEGRATION

3-101. Before observing a subordinate, leaders should review the performance indicators (see chapter 6) to associate observations with the various levels of proficiency under each competency and attribute. Before the observed event, leaders should record the situation and include information such as the date, time, place, and mission or task the subordinate is involved in. They should also note any other key players in the situation and the climate of their relationship (if known).

3-102. During and immediately following an event, leaders should record their observations. Referring to the behavioral indicators to associate observed behaviors with the competencies and attributes, leaders should indicate proficiency as either a developmental need, meeting the standard, or a strength.

3-103. Following the observation event, supervisors should record how to reinforce the observed behaviors and note recommendations. When delivering these observations to the subordinate, the leader should refer to the recorded notes. During delivery, the leader should be prepared to highlight the subordinate's strengths, how they meet the standard, and developmental needs. The discussion should lead to reinforcement and recommendations for sustainment and improvement.

3-104. Leaders should engage in a guided discovery learning conversation with subordinates by asking open-ended questions to help them understand the effect their actions had on the mission and Soldier outcomes. Leaders should guide the subordinate toward the realization of strengths and improvement of developmental needs.

OBSERVATION STEPS REVIEW

As a review of the process for delivering an observation, these steps help leaders deliver observations to subordinates—

Confirm the situation.

Ask for a self-assessment.

Clarify and come to an agreement.

Add observations.

Raise future-oriented questions; ask for recommendations.

Strengthen the leader—validate and reinforce positives.

COACHING

3-105. Coaching helps another individual or team through a set of tasks or with improving personal qualities. A coach gets the person or team to understand their current level of performance and guides their performance to the next level. A central task of coaching is to link feedback interpretation with developmental actions. The role of the coach is to advise the individual or team on what levels can be reached and what to do to reach them.

3-106. Similar to other development processes, there are a number of components in coaching:

- Building rapport. The coach builds a strong rapport to facilitate trust and open communications.
- Gathering and analyzing information. Performance indicators or the leader or team's perceptions are reviewed to determine an accurate picture of capabilities.
- Addressing the gaps. Specific issues are discussed in light of similarities and differences with what are normal expectations.
- Narrowing focus. The coach helps guide the leader to identify the directions to strengthen and develop.
- Setting goals. The coach assists the leader in establishing development goals.
- Planning development. Together the coach and leader determine paths of development, desired outcomes, and specific developmental actions.
- Promoting action. The coach sets conditions that help to sustain developmental action and establish accountability for development.

3-107. Coaches can draw on the guided discovery learning techniques to establish and maintain rapport and to build commitment. The coach tailors how directive feedback and guidance are depending on the situation of those being coached and the performance level. If coaches are involved in developmental actions, they look for a good balance between challenge and the learner's perception of ability to achieve incremental improvement.

3-108. To prepare for coaching, leaders will study and apply the fundamental guidelines for leader development. They will be passionate learners in the area being coached. They will arm themselves with tips, techniques, and practice routines to advise subordinates. Developmental actions for leadership include observing other leaders, modeling what good leaders do, and practicing new techniques or approaches. Leaders can apply techniques in the conduct of their duties, look for different on-the-job opportunities, or identify outside opportunities. Other actions include reading, research, consulting, and formal coursework. Sometimes applying different mindsets and ways of thinking provide enough development to meet established goals. ADRP 6-22 provides guidelines for coaching (focus goals, clarify self-awareness, uncover potential, eliminate obstacles, develop action plans and commitment, and conducts follow-up).

STUDY

3-109. Leader development processes in the organization should establish an expectation for each leader to spend personal time seeking sources of knowledge and opportunities to grow and learn. If a supervisor's personal involvement and unit resources were always prerequisites for leader development, it would be a limited effort indeed. Organization leaders should develop distinct ways of studying their chosen profession and identifying ways to improve the unit.

Encourage subordinates of the same position or similar grade to form a community-of-practice group that fosters excellence. Provide the groups reachback capability to Web-based forums. Provide each group with an opportunity to present recommendations or new methods to the leadership team.

Professional Reading Programs

3-110. Professional reading programs broaden leader knowledge, understanding, and confidence. Leaders gain a refined understanding of the material and develop critical thinking skills through pertinent discussion with others. Discussing ideas and topics with peers, subordinates, and leaders who may offer significantly different perspectives exposes all participants to new ideas and potentially broadens their outlook.

Successful reading programs depend on how they are structured—what readings are chosen and what purpose is integrated into the program. If you want to encourage tactics, then select readings on operational tactics. If you want to develop skills for which interesting readings do not exist, then design questions that trigger reflection about engaging material. For example, to stimulate critical thinking assign questions about the materials that require consideration of underlying assumptions, alternative courses of action, and application of lessons to other situations.

3-111. Organizations and individuals can implement professional reading programs; a wealth of materials are available to support topic determination, such as the U.S. Army Chief of Staff's Professional Reading list or the U.S. Center of Military History Professional Reading List. Determining the frequency, such as monthly or quarterly, will be dependent on organizational missions, but the unit must allocate and protect time for effective implementation.

3-112. For personal professional reading, topics may come from established reading lists, stem from personal interests, or follow from determining strengths and developmental needs. As part of a personal reading program, leaders should maintain a reading journal to take notes and record key passages, insights, and reflections. Leaders who record thoughts on paper can gain clarity and develop new ideas. The journal could record titles of related books and articles for further investigation.

PROFESSIONAL WRITING PROGRAMS

3-113. Army leaders consider how they can contribute to the body of thought in their fields of expertise by researching and writing about topics that interest them. By writing and publishing papers, they can advance their profession, their mastery of their discipline, and their writing skills. Writers of scholarly papers study their topics in depth and in breadth. They take formal classes in research and in writing so they can master appropriate standards. They use appropriate writing processes. Before submitting papers to professional or academic journals, they ensure their submissions meet the publications' requirements. In addition, the unit security office should screen items for publication to prevent the spillage of classified information. Writers scrupulously adhere to intellectual property rights rules and shun plagiarism.

3-114. For developing leaders, a developmental writing program serves as a significant complementary companion to a professional reading program. Length and time given for completion should vary based on the requirement. Some ideas and suggested lengths for professional writing include:

- Leadership philosophy—an opportunity to codify what you believe as a leader such as expectations, what is important, and what is non-negotiable (2-3 pages).
- Personal experiences:
  - Significant experience, whether good or bad, and how it affected you including lessons learned (5-7 pages).
  - Routine experiences, describing how you handle them and possible improvements for consideration (2-3 pages).
- Historical person or event related to your branch, regimental affiliation, or organization (5-7 pages).
- Opinion piece explaining changes affecting your branch through a particular person, policy, or equipment (5-7 pages).

3-115. Individuals should consider writing for publication as a complementary element to the professional reading program. Writing increases self-development as well as develops others who gain from the lessons learned and stimulated thought. Papers created through the writing program could be considered for publication in branch journals or as blog entries. This shares ideas and gathers feedback for the author, which could be beneficial in further developing the original ideas.

CONSIDERATIONS FOR PROFESSIONAL READING

A reading program is one way to foster self-study by making the reading relevant, provide a purpose, and follow up. Leaders can use this format to present a short lesson on leadership and leader development to others.

Book/Article/Reference
Name of leader and position
Describe the leader's environment and situation.
Who was the leader leading?
How did the leader attempt to influence the situation or people?
What were the positive and negative outcomes?
What were the leader's strengths and development needs?
What lessons from this leader's experience can we apply immediately or later?

Additional questions may be used to focus readers on specific aspects:
What is the significance of the title? Would you have given the work a different title? If yes, what is your title?
What were the central themes? Do you feel they were adequately explored? Were they presented in a cliché or in a unique manner?
What did you think of the structure and style of the writing?
What part was the most central to the work?
What resonated positively or negatively with you personally? Why?
Has anything ever happened to you like the examples cited? How did you react?
What surprised you the most?
What were notable historical, economic, racial, cultural, traditional, gender, sexual or socioeconomic factors brought up in the book? How did they affect the presentation of the central idea? Was it realistic?
Were there any particular quotes that stood out? Why?
Did any of the situations remind you of yourself or someone you know?
Did you disagree with the author's views? If so, what specifically and why?
Have any of your views or thoughts changed after reading this?
Are there any works that you would compare with this one? How does this compare?
Have you read any other works by this author? Were they comparable to this one?
What did you learn from, take away from, or get out of this work?
Did your opinion change as you read it? How?
Would you recommend a peer, subordinate, or supervisor read it?
Was anything confusing or contradictory about what the author presented?
Why do you think the author included some of the stories?
What is your rating for the work? How do you feel about reading it?

SECTION IV – CREATING OPPORTUNITIES

3-116. Creating opportunities for development or using existing experience opportunities is a fourth way of creating a culture of development. An organizational culture develops based on shared values, beliefs, and learning. These cultural values, when consistent with the mission, affect an organization's performance. Leaders foster a positive culture by providing a supportive command climate that values member involvement and learning. Likewise, the selections for and responsibilities of key positions of leadership will have implications for developing leaders far into the future. Integrating these efforts into a holistic program will establish lasting operating norms. Developing leaders to this level requires an investment of time and effort, but leaves a lasting legacy of trained and ready leaders for the Army of tomorrow.

3-117. Selection and screening of leaders can be useful in leader development efforts. Forming leadership teams where strengths in one complement developmental needs in another is a common selection goal. Developing leaders is often about preparing them for responsibilities in the next position. Creating opportunities for leader development involves—

- Creating challenging experiences.
- Sharpening leader selection.
- Planning leader succession.
- Tracking career development and management.

CHALLENGING EXPERIENCES

3-118. Experience is a developmental tool. leaders can create learning opportunities by placing subordinates into challenging assignments to stretch their thinking and behavior. Challenging experiences are characterized by pressure, complexity, novelty, and uncertainty. Challenge creates learning situations that are interesting and motivating. Leaders can also create these experiences or ensure opportunities are used as learning experiences.

3-119. All Army assignments inherently provide a degree of developmental challenge. Leader development will happen even if supervisors do nothing at all. Creating the right challenges in a position for a particular leader can dramatically increase development.

3-120. Some missions or circumstances may not offer key developmental opportunities. Supervisors may need to shape position responsibilities to allow a subordinate to enhance personal leadership skills. Before adjusting a position's requirements, leaders should consider unit and mission demands.

3-121. Leaders should be deliberate placing subordinates in special missions and organizational assignments. Experienced leaders implicitly know the defining tasks early in an assignment and should be deliberate about identifying these tasks and ensuring each leader gains experience from them. Sometimes, supervisors must assign subordinates to positions for which they do not have the requisite skills or experience. Supervisors should consider modification of position requirements and providing additional support or resources.

3-122. Not all leaders develop on the same timelines. Supervising leaders should be willing to adjust how much time each subordinate stays in a position. Supervisors should involve human resources staff early in these discussions as decisions may have implications beyond the organization. When making such determinations, supervisors should weigh the effects on—

- Unit performance.
- Stability of the leadership team.
- Leadership needs of adjacent units, higher units, and the Army.
- The leader's well-being and personal growth.

In determining what subordinates need to learn, ask them about the top three skills they need to become proficient to improve unit performance. Doing so will motivate them and increase their reception to the leader skills they need to learn.

Designing Assignments for Development

From a battalion commander:

I led a group of company commanders with very different communication styles, interpersonal skill levels, and backgrounds. I perceived each needed some individual and special experiences to better round out their skill sets. Therefore, I purposefully assigned projects that put them each in challenging situations requiring them to prepare for and execute with skills that were not their natural strengths, but nonetheless important. For example, I would take someone I considered to be introverted and have them meet with the family readiness group, which I know was likely be comprised of a lot of diverse folks and require effective communications and relationship building. That task provides a junior leader with a different environment than their direct chain-of-command and consequently expands their communication and interpersonal skills. I think they learn more from some of these types of tasks than just daily Soldier interaction tasks.

The goal in all of these day-to-day task assignments was to get leaders out of their comfort zone, and away from isolating themselves and stovepiping their actions within a narrowly defined, known environment. I also wanted to broaden and build their skills sets and adaptability. I did not necessarily expect everyone to become fully "rounded," but I did think it encouraged and resulted in individual growth. It definitely improved my company commanders' ability to deal with change and novel situations.

LEADER SELECTION

3-123. Supervising leaders should foster an attitude that leadership positions are not necessarily automatic appointments. It is a privilege, not an entitlement, to serve in a leadership position. Selections for key leadership positions require thorough consideration. Each step in the screening and selection process should narrow the field of acceptable candidates. For key leadership positions, a deliberate selection process should be followed:

- Forecast potential position openings.
- Identify key leader characteristics.
- Build a pool of candidates by working with higher, adjacent, and subordinate units, as applicable.
- Use selection tools to screen out applicants such as—
  - Conducting a career file review to identify prerequisite experiences and training; review files and rate candidates against career indicators.
  - Reviewing disciplinary or derogatory information in personnel and intelligence files.
  - Obtaining references or recommendations on the leader from trusted sources.
  - Conducting structured interviews with candidates for the position—structure the interviews to assess values, attributes, and responses to various situations.
  - Organizations may develop minimum prerequisite knowledge or skills requirements for particular positions. Final candidates may demonstrate capabilities by conducting a task that proves their qualifications for the position (such as leading a patrol or leading a convoy).
- Select and appoint approved candidates.

3-124. If creating a pool of qualified candidates is not possible, supervising leaders should consider modifying the position or providing additional support or resources to available candidates.

3-125. These processes fall within the realm of talent management, which complements leader development. Talent management takes into account the individual talents of an officer, NCO, or Army Civilian—the unique distribution of personal skills, knowledge, abilities, and behaviors and the potential they represent. The Army looks to develop and put to best use well-rounded leaders based on the talents they possess—talents that derive not only from operational experience but also from broadening assignments, advanced civil schooling and professional military education, and demonstrated interests.

Consider the leadership team when selecting leaders. For example, pair a technically strong warrant officer with a tactically strong officer. Pair a strong operations officer with an intelligence officer willing to challenge the operational plan by forcefully presenting the enemy point of view. Pair an experienced NCO with an inexperienced lieutenant.

LEADER SUCCESSION

3-126. Succession planning is a developmental activity for the individual leaders that focus on deliberate planning to provide opportunities for experience in key developmental assignments and to prepare for future assignments beyond the unit. Unit leaders do not have total input into succession planning but with forethought can have plans developed to rotate leaders within the unit. Succession planning is a localized version of talent management. Senior leaders plan the systematic rotation of subordinates within the organization so that trained and qualified leaders are ready to assume vacancies, proven leaders move on to positions of greater responsibility, and marginal leaders receive opportunities to improve. Succession planning serves individual leaders by looking beyond the replacement interests of the organization. It helps develop leaders with the potential to succeed in future positions beyond their current unit and returns a benefit to the Army by optimizing development opportunities and duration across the unit's leaders.

Grant and McPherson

During the American Civil War, Ulysses S. Grant rose to become the Commander in Chief of the Union Army. Along the way, he groomed a select number of officers to succeed him. Those he supported for further advancement showed three attributes: personal loyalty, a willingness to do any duty necessary to prosecute the war, and a desire to prove oneself in battle. One of Grant's inner circle who gained his full trust and confidence was James B. McPherson.

McPherson was a Regular Army officer who graduated from West Point in 1853 and commissioned as an Engineer. Eager to find a combat assignment, he joined Grant's staff in January 1862 after promotion to lieutenant colonel. McPherson served admirably as Grant's chief engineer during the Fort Donelson campaign and at Shiloh. A rising star, Grant promoted McPherson to major general in 1862 and appointed him to command an infantry corps. His successes during the Vicksburg campaign cemented his reputation. When Grant was promoted and sent east, he designated McPherson as commander of the Army of the Tennessee—a wise choice.

Grant knew the old army adage that "best friends may not always make the best generals." McPherson had indeed become Grant's friend over the years. However, McPherson's ability to see Grant's goals and work tirelessly to meet them won the full confidence and support of his commander. McPherson did not disappoint. While other generals sought to seize ground and take cities, he endeavored to engage and destroy enemy armies. McPherson's army was successful in driving the Confederates back through northwest Georgia as a part of Sherman's Atlanta campaign. Leading from the front, McPherson was killed in action on 22 July 1864. Grant memorialized McPherson as one of the "ablest, purest, and best generals." Sherman called him "a man who was...qualified to heal national strife." Even his adversary, John Bell Hood, marked his passing with friendship and admiration. Yet while he lived, McPherson proved a sterling example of how to establish a succession of command.

3-127. Understanding the projected career paths and timing for leader branches and specialties is an important factor in succession planning. Moving leaders into and out of positions should be a factor of—

- Unit performance. Keeping leaders in positions long enough so that their stability promotes high unit performance.
- Army need. Providing experienced leadership back to the Army to fulfill its requirements.
- Individual leader developmental goals and readiness. Determining when the leader has achieved development goals and is ready to take on new responsibilities and challenges.

3-128. Supervising leaders should work with human resources staff to predict accurate leader gains and losses to the unit. Be sure to—

- Account for leader needs for career and position-specific training before position assumption.
- Assess leaders during their initial assignments to drive subsequent position assignments.
- Use leader vacancies due to schooling, special assignment, or leave as leader development opportunities; assign less experienced leaders temporarily to the vacancies.

Identify the key leadership positions that trigger succession planning and management. Chart the timing and sequencing of leaders into and out of unit leadership positions. Account for prerequisite schooling and plan primary and alternate candidates for each position.

Be an Advocate for Yourself—Take a Career-view

From a battalion commander:

One of my prior battalion commanders once told me, "When you're working one position, know the next three you want down the road, and focus on how to get there to ensure you attain that goal. Always keep thinking about the next three jobs you want to do. When you're in one job, always do that job, but also be training yourself to do that next." I had never thought about that before. My commander shared that with me while we were in Korea. He was great about day-to-day interaction and passing on professional advice and lessons learned from his experience. His technique was to visit with us often and unit performance was always his first priority and order of business. Once he got a sense of what was going on with the unit, he would motion for me to follow him back to his vehicle, and it was then that the one-on-one sharing would start. It could be a lesson learned or a story about how he handled a situation, or sometimes just asking me about how I was doing. I did not fully appreciate what he was doing then, but I do now. He helped me not only with the work I was doing, but also thinking ahead both professionally and personally.

Eight years later, I still use much of what he shared with me. Whenever I talk to my leaders and they ask me about my goals, I am prepared. Recently, the commanding general asked me, "Hey, what do you want to do?" Because of what I learned from my commander in Korea, I responded, "Hey sir, here's who I am. This is what I need to do to get the jobs I want. Here are my goals for those jobs. Here are three jobs I want and three different timelines for how I can achieve them." Then your leader can take that all in and help you make some decisions. For instance, the general told me, "Your course of action #2 is the most realistic. It's probably best suited for you and will help you in your career progression."

CAREER DEVELOPMENT AND MANAGEMENT

3-129. Individuals should understand and actively manage their own career paths while supervisors should consider the career paths and influence their subordinates to gain breadth in development. Commanders and other senior leaders should encourage their developing subordinates to take challenging assignments. Reserve Component leaders should be aware of subordinates' civilian development plan as this may affect their ability to take on new and challenging assignments.

3-130. The Army provides ACT and other online tools to help leaders in collaborating with their subordinates in professional development planning discussions. Supervisors must provide opportunities for subordinate's personal and required individual learning. ACT enhances personnel counseling by providing a framework to create IDPs and the ability to monitor career development while allowing leaders to track and advise subordinates on personalized leadership development.

Balance of Army Needs with Personal Choices

3-131. The gravity of the Army mission and the dynamic nature of the world make continuous learning and self-development crucial to personal success and national security. Rapid changes in geopolitical affairs, technology, and general knowledge require individuals to repetitively seek current information. Army and civilian schools provide basic knowledge, frameworks, and techniques that individuals need to continue to review and update after they leave those schools. To thrive professionally and personally, individuals must engage in life-long learning and self-development.

3-132. Finding the proper balance between professional work and personal life while planning career development challenges professionals at all stages of their careers. Most career planning models have the following common steps—

- Perform a self-assessment to determine strengths and developmental needs (based on abilities, characteristics, needs, responsibilities, or interest or goals).
- Weigh the possibilities to choose goals and milestones for self-development efforts.
- Make a self-development plan that uses effective methods of learning.
- Implement the plan, overcoming obstacles, and measuring progress.

Ask the organizational leaders to describe their most valuable leader development experience. Give them a few days to think about it before they respond. Have them briefly write the experience down or tell it to a group of their peers. Use their experiences to help prioritize implementation.

Providing a first, second, third, and fourth priority reflects the understanding that leaders may not be able to implement every idea or method. Some methods of leader development provide a leader with a higher return in performance for less investment of resources.

Civilian Training and Development Programs

3-133. The Civilian Education System is a structured program with central funding for all Army Civilian personnel and serves as the foundation for civilian leader development. Army Civilians have developmental opportunities afforded by their duty series, grouped into career programs. Each career program makes available career planning tools to enable the development of core competencies. ACTEDS provides a planned course of professional development, using a combination of formal training and education and progressively challenging work experiences. ACTEDS is a resource for both individuals and for their supervisors. Through ACTEDS, Army Civilians have the opportunity to plan and conduct their own development.

3-134. Information is available for each Army Civilian career program at Civilian Personnel Online and ACT. For example, Career Program 34 is a 14,000-member group of Information Technology Management personnel. Development programs and opportunities for Career Program 34 include academic degree training in such areas as technology management, information technology management, information security, and computer science. Short–term training is available in areas such as project management, cloud computing, system administration, and software development. Some opportunities are competitive and slots are filled through application, nomination and screening of candidates. Management programs are available from Office of Personnel Management development centers, executive leadership development, and executive potential programs.

3-135. Developmental assignments are encouraged to broaden knowledge of how different organizations conduct information technology and cyber missions. Training with industry is available to higher-ranking personnel where they learn about information technology practices outside the Army. Distributed learning resources are extensive for the Career Program 34 population, predominantly through Army e-Learning courses and certifications. Each career program has similar opportunities to guide the professional development of Army Civilians.

> **Team building as Development**
>
> From an Army Civilian supervisor:
> Team building projects are a huge success. Together we have a brain storming activity or a process review. We discuss where the team feels shortcomings. We have improved directorate communications, eased document processing, made procedures more efficient, and clarified desired outcomes from a team perspective. We also do monthly training and pass the helm to someone different every month to run it. This gives them a sense of responsibility, helps them feel engaged, and affirms their expertise. It develops their self-confidence as well.

Professional Development Programs

3-136. Leader professional development programs bring an organization's leaders together for a specific developmental purpose. Leader development programs are an effective vehicle for leader development when consistently applied. Common elements of successful programs include—

- Mission-essential leader task training when a common need exists across the organization.
- Required orientation or education sessions (such as equal opportunity and safety).
- Cohesion-building activities that foster esprit de corps (such as a dining-in, sports, or adventure training).
- Opportunities for the commander, command sergeant major, or first sergeant to emphasize key guidance to all leaders.
- Education sessions on leader career path topics (assignments, schooling, or promotions).
- Education sessions on the mission command philosophy, culture, and geopolitical issues.

3-137. Professional development sessions, conducted to facilitate discussion and collaboration, are extremely valuable in gaining a greater understanding and application of specific information or skills in a unit. The sources of information and means of conducting these sessions are endless and allow for creativity. Instructors should not rely solely on dry briefings. Scenarios and materials should be tailored to the grades and ranks present. These sessions can be great team building opportunities to bring together groups of different ranks and responsibilities.

3-138. All of these applications fulfill the training and development needs of the leaders in the organization. To implement leader development programs effectively, leaders should invoke the following guidelines—

- Link training and professional development.
- Clearly communicate purpose and relevance.
- Gather all leaders together only when doing so is the most effective learning method.
- Consider prior listed applications as integral to leader development programs.

To provide leaders with an in-depth perspective on a mission-essential task for the organization (such as security patrols or convoy operations), supervisory leaders should lead the task while subordinate leaders perform the roles of Soldiers. By practicing the execution of the task to standard, the organization's leadership will be more effective at supervising future execution.

DEVELOPING MISSION COMMAND

This exercise offers a technique to understand and create a culture of mission command through subordinate participation.

Step 1: Educate the unit on the philosophy of mission command. Everyone down to the lowest-ranking member needs to understand the mission command philosophy, why it is needed, and what it is not.

Step 2: Every member of the organization has an assignment to imagine that overnight the unit wholeheartedly embraced the philosophy of mission command with everyone coming to work the next day acting and behaving differently. Based on that premise, they answer the following:

What would counseling look like for the organization?
What would my responsibility be?

What would morning physical training look like for the organization?
What would my responsibility be?

What would training events look like for the organization?
What would my responsibility be?

What would our organizational leader development program look like?

What would my self-development program look like?

Step 3: Use the responses to develop clearly articulated goals and behaviors to provide guidelines and visible markers that the exercise of mission command actually happens. For example, several comments recommend quarterly developmental counseling for everyone with no counseling completed with a generic fill-in-the-blank counseling form. Follow the survey with professional development classes by leaders (grade immaterial) known for excellent developmental counseling sessions. Make this an organizational goal to accomplish for the next quarter.

Opportunities during Training Events

3-139. Training is an organized, structured, continuous, and progressive process based on sound principles of learning designed to increase the capability of individuals, units, and organizations to perform specified tasks or skills. The objective of training is to increase the ability of competent leaders to perform in a variety of training and operational situations. Individual task training builds individual competence and confidence to perform these tasks to support collective training and operations.

3-140. Leaders contribute substantially to the unit's mission success or lack of success. Therefore, the Army devotes considerable resources to foster leader development during exercises. Leader development is an important duty of supervisory leaders and the leader's chain-of-command. Their responsibility is to provide leaders with accurate observations of their abilities and the effects on unit performance. Providing leadership feedback is a difficult, yet essential, part of training exercises. Without it, the assessment of an important contributor to a unit's mission accomplishment, namely its leadership, is left undone.

3-141. Leaders have a specific task to observe subordinates during planning and executing missions. Some may feel unqualified to observe and provide feedback on leadership actions. However, understanding how to treat leadership as a set of skills that can be developed and improved is essential.

3-142. Guided discovery learning is an important underpinning of developing leaders. To the extent possible, leaders ought to use guided discovery learning. Doing so places the observed leader in charge of their learning, with the chain of command in a supporting role. Using guided discovery learning during training exercises makes the leader better prepared to be a self-guided learner in any contemporary operational environment. Providing feedback falls in the larger context of guided discovery learning methods. Chapter 6 provides leader performance descriptions at various levels of proficiency to support leader observation and feedback.

Integrating Development into Daily Events

From a battalion commander:

You deal with all types of Soldiers' issues as a commander and command is fundamentally about motivating and influencing Soldiers. Therefore, I mentor my junior officers by preparing them to look at Soldier issues from a command perspective. I integrate this daily activity into my interactions with subordinates and field visits. For example, whatever position a junior leader happens to be in at the time, I mentor them by asking them how they might deal with the issue if they were the commander. I will let them offer an idea and then pose a few questions to help them gain from my experience. I find this informal but deliberate mentoring has the benefit of reinforcing my commander's intent through the ranks. Everyone starts thinking like a commander and taking ownership of issues at their level. That leaves me time to step back and command. Experience with Soldiers and understanding the human dynamic is what makes a successful commander. Sure, you have to have general knowledge of what the equipment is and how to employ it. However, it is how you as the leader make Soldiers proud to go to war and take care of their families and develop themselves into better leaders and people.

Chapter 4
Self-development

4-1. Self-development bridges the gaps between the operational and institutional domains and sets the conditions for continuous learning and growth. Military and Army Civilian personnel engage in self-development to improve their capabilities for current and future positions. Self-knowledge is an important part of a leader's development. Several tools, such as the Army MSAF program, are available to leaders to understand strengths and obtain insights into developmental needs.

4-2. Self-development is an individual's responsibility but it is important for leaders to set conditions and support self-development. Leaders need to be actively involved in developing themselves and each other. Development happens through study and practice. Leaders can support others' self-development through the exchange of professional development information, discoveries, and opinions.

4-3. Self-development supports planned, goal-oriented learning to reinforce and expand the depth and breadth of what a person knows to include themselves and situations they experience and how they perform their duties. The Army acknowledges three types of self-development:

- Structured self-development includes mandatory learning modules required to meet specific learning objectives and requirements.
- Guided self-development is recommended, optional learning intended to enhance professional competence.
- Personal self-development is self-initiated learning to meet personal objectives such as pursuing a college education or an advanced degree.

4-4. To help subordinates learn from their experiences, leaders should provide opportunities for them to pause, reflect, and process the experience for what was learned. Reflecting on an experience—

- Keeps leaders from repeating the same mistakes.
- Helps leaders consider effects in future decisionmaking.
- Helps leaders to link their actions with the resulting effects on unit performance.

Working environments can be chaotic, noisy, and filled with activity. However, prioritizing time for reflection and consolidation of thoughts enhances self-development.

4-5. The self-development process consists of four major phases. They are—

- Strengths and developmental needs determination.
- Goal setting.
- Self-enhanced learning.
- Learning in action.

STRENGTHS AND DEVELOPMENTAL NEEDS DETERMINATION

4-6. The first step in determining strengths and developmental needs is to think about what you do and how well you do it. At a minimum, this information comes from self-examination. Outside opinions and information on strengths and developmental needs are useful. Feedback can come from formal or informal assessments and from other leaders, peers, or subordinates. Keep this in mind during a self-examination.

4-7. Understanding current strengths and developmental needs is necessary before setting self-development goals. This is part of being self-aware. These methods help identify strengths and developmental needs:

- Information collection.
- Feedback gathering.

- Self-analysis.
- Strengths and developmental needs identification.

INFORMATION COLLECTION

4-8. Formal assessments such as evaluations and tests are a good place to start gaining insight into strengths and developmental needs, since they measure individual performance and compare it to a standard. Individuals use the information and results from relevant assessments to inform understanding of personal strengths and developmental needs. Formal assessments include—

- Performance evaluations.
- Counseling sessions (formal and informal).
- Skills tests (such as the Expert Field Medical Badge and Expert Infantryman Badge tests).
- Tests administered in resident and non-resident schools.
- Field performance evaluations such as those at the combat training centers.
- Intelligence and aptitude tests (such as Armed Services Vocational Aptitude Battery or Defense Language Aptitude Battery).
- MSAF program feedback (360-degree assessment) where superiors, peers, and subordinates provide anonymous feedback.
- Occupational interest inventories.

Unique Value of 360 Assessment

From a first sergeant:
Nobody ever tells you if you are screwed up. They just talk behind your back or assign the mission to someone else. With the 360 assessment, you get the truth. This is not just another survey; it is essential.

FEEDBACK GATHERING

4-9. Hearing what peers, subordinates, superiors, mentors, family, and friends think can help identify strengths and developmental needs that went unnoticed or you have been reluctant to acknowledge. There are two ways to get feedback: observe how others interact with you or ask them directly. Supervisors have an explicit role in subordinate development. Subordinates should consult supervisors for guidance about development goals or any other aspect of self-development.

OBSERVE OTHERS

Observing how others act toward you and the decisions they make affecting you will give an idea of what they think about your skills and expertise. When observing others:

Make observations on different occasions. Watching the same person several times will help you see trends that may be a sign of a firmly held opinion of you. One observation is not reliable, as that behavior may have been a result of other issues.

Consider the circumstances. What outside factors influenced the person's decisions and actions? For example, if your supervisor selected someone else to perform an important task, was it because you were too busy or unavailable?

4-10. After considering these questions, analyze the answers to determine the opinions that each person considered may have of your strengths and developmental needs:

- Supervisors, raters, and superiors.
  - Who gets the most challenging assignments?
  - The supervisor relies upon whom during emergencies or tough problems?
  - The supervisor praises whom the most?
  - What kinds of tasks does your supervisor give you versus others?
  - How does your supervisor react to your suggestions compared to others' suggestions?
  - Does your supervisor listen to your opinions on certain subjects much more or much less than the opinions of others? If so, what are those subjects?
- Peers and Subordinates.
  - Do peers and subordinates come to you for help or advice? In what topics?
  - Do they understand you or seem confused or overwhelmed by what you say?
  - Do they repeatedly contact you for help or are they one-time interactions?
  - Does their interest and enthusiasm increase or diminish when they interact with you?
  - What does their body language communicate? Is it relaxed, apprehensive, or reserved?

Asking for Feedback

4-11. One can learn a lot about others' perceptions by observing interactions, but conclusions will only be educated guesses unless the observers are asked directly. To gain perspective, talk to others who know you in different ways, such as one's rater, enlisted or officer counter-part, mentor, instructor, or family member. The goal is to find out—

- What a person actually saw and their impressions of your action(s).
- That person's impression of how well you performed during the interaction(s).
- How you react in certain situations. For example, "When a subordinate challenges your authority in front of others, you seem to get flustered and be at a loss for words."

Who to Ask

4-12. These are items to consider when determining who to ask for feedback—

- Who has been able to observe you enough to offer useful information?
- Who has observed you from different perspectives?
- Who has experience in an area of interest (former or current supervisor, mentor, or teacher)?

Things to Remember When Asking for Feedback

4-13. When asking for feedback, keep the following in mind—

- Be respectful of others' time—prepare questions ahead of time.
- Approach with an open mind to accept uncomfortable or critical feedback without offense.
- Listen carefully and respectfully.
- Ask for clarification and examples when points are unclear.
- Summarize the points to make sure that you understand the person correctly.
- Thank the feedback providers for their time and assistance.

4-14. These ideas may help you focus on what to ask:

- Get descriptions of your behaviors and opinions of those behaviors.
- For feedback about a recurring issue, ask about the situation, your actions, and the usual outcomes.
- Ask for suggestions for other ways of handling situations.

4-15. Compare the feedback received from different sources to look for common themes. These themes will help to identify strengths and developmental needs. Army leaders must try to avoid the natural inclination to reject or minimize responses that do not confirm self-perceptions or attribute them to the situation instead.

Self-Analysis

4-16. After gathering the information from outside sources through formal assessments, observing others, and requesting feedback, it is time to reflect on personal behavior and performance. Examining personal situations and experiences can reveal things to change or improve. The situation analysis exercise will help analyze experiences to help identify personal strengths and developmental needs.

COMPLETE A SITUATION ANALYSIS FOR SELF-DEVELOPMENT

Think of experiences over the past two years that give insight into personal strengths and developmental needs—maybe a critical decision, an important task you led or were a part of, or a significant personal interaction. Use these questions to analyze each situation:

What was the situation? What was happening? Who was there?

What was the goal and did you reach it? What were you trying to accomplish? What resources or skills did you have or not have that you needed?

What did you say and think? Were you able to find the right words to make your point? What were you thinking at the time? What made you feel good (confident, excited) or bad (confused, worried)?

What did you do? How did you act (including your body language)? Why did you choose to act the way you did? How did others react? Did you help or hurt the situation? Did you adjust your actions based on how others were reacting?

Why did you act the way you did? What knowledge and skills led you to act the way you did?

What could have helped you handle the situation better? How could you have used your strengths to reach a better outcome? Are there any developmental needs that you should make a high priority for personal self-development efforts?

4-17. After recording the information, look for key factors that influenced the situation progression and the overall outcome. Keep in mind that if the same factor occurs in multiple situations, it may suggest a significant strength or developmental need that may be developed.

4-18. By knowing how personal actions affected the situation and the thoughts and feelings associated with those actions, leaders can work to become more self-aware and choose the most productive actions. In addition, a self-analysis may suggest broader interests to pursue or issues to avoid.

COMPLETE A SELF-ANALYSIS

Consider the following items and be as specific as possible. Use the items as necessary to identify unique aspects of personal strengths and developmental needs.

Strengths
The skill or ability at which I am best is—
The personal quality that I rely on most for my success is—
I am most knowledgeable about—
The activities I look forward to include—
I would love to learn more about—
The accomplishment I am most proud of is—
Others usually come to me for help with—
Others think the best position for me would be—

Developmental needs
The skill or ability that is always difficult for me is—
I don't know as much as I should about—
I usually go to others for help on—
The situation that causes me the most frustration is—
I am most hesitant when I try to—
I am most concerned about my—
Others think I am not skilled at—
I would become a more valued member of my organization if I—

STRENGTHS AND DEVELOPMENTAL NEEDS IDENTIFICATION

4-19. The final step is to take the information gathered from formal assessments, information gathered from observing others and asking others, and results of the situation and self-analyses and analyze it to determine strengths and developmental needs.

4-20. Instead of taking all of the feedback as fact, look for recurring themes or patterns of feedback heard from more than one person. Look at what others identified as strengths and developmental needs and compare that to personal knowledge (from the self-exam) and the results of formal assessments.

4-21. Usually, repeated success or expertise in a particular activity indicates a strength. These abilities may come easily even though others find them difficult:

- What are your favorite things to do or learn about?
- What do others turn to you for help with?
- What do recent assignments show as strengths?

4-22. Developmental needs are tasks that are a struggle to learn or difficult to perform:

- What was noted as being hard or not fun to do?
- What did others suggest as a limitation?
- Did formal assessments point out any deficiencies?

4-23. Identify where these descriptions apply and make a list of strengths and developmental needs. This list will enable setting clear goals for self-development efforts.

GOAL SETTING

4-24. To make the most of self-development efforts and avoid wasting time and energy, it is crucial to set self-development goals—identify personal and professional goals and decide where to go. This section outlines procedures to—

- Gather the information needed to decide how to structure self-development efforts.
- Establish self-development goals.
- Plan milestones to keep on track.

Information Gathering

4-25. An understanding of strengths and developmental needs is an important place to start when determining where to focus self-development efforts. Other areas to analyze for self-development opportunities include—

- Roles and responsibilities (personal and work-related).
- The needs of the Army.

Roles and Responsibilities

4-26. Roles and responsibilities at home and at work may offer opportunities for self-development, such as being a spouse, parent, teacher, Soldier, or other roles. Each role has different responsibilities, skill and knowledge requirements, and expectations. Reserve Component leaders have a unique opportunity to improve both civilian and military profession skills by linking self-development goals to skills shared by both professions. ADRP 6-22 describes expectations for key roles as Army military and civilian leaders.

4-27. Chosen roles usually reflect personal interests and values, but even assigned roles will affect the value of different self-development paths. When roles and responsibilities align with talents and interests, leaders are likely to succeed and be satisfied.

ANALYZE ROLES AND RESPONSIBILITIES

List three to four of important roles at home and at work. Next to each role, list the two most important responsibilities in that role.

Now think about the listed roles and responsibilities and identify knowledge, skills, or attitudes that enable better performance of these roles and responsibilities.

Needs of the Army

4-28. Another way to identify satisfying goals for personal self-development efforts is to align personal interests with Army needs. This ensures that the acquired knowledge and skills are personally interesting but benefit the Army.

4-29. Soldiers and Army units must be ready to deploy to any part of the world and accomplish diverse missions. Some requirements may be unforeseen and untrained, requiring Soldiers to use their knowledge, skills, and creativity to accomplish the mission. As members of a unit develop expertise in a variety of areas, the unit and the Army as a whole become stronger. The range and depth of expertise gives the unit resiliency and an increased ability to adapt to specific challenges.

> **General Joe Stilwell's Commitment to Self-Development**
>
> General Joseph Stilwell served as the commanding general of U.S. forces in the China-Burma-India Theater during World War II. He rose to that position through a career focused on developing his understanding of the Chinese language and the Chinese people. Stilwell gained much of this knowledge through persistent pursuit of personal study and development. As a junior officer, Stilwell found he had a gift for languages and constantly sought opportunities to develop his ability to speak other languages and understand foreign cultures. Before World War I, he used personal leave to journey extensively through Latin America and parts of Asia, perfecting his Spanish and picking up basic Japanese on these travels. Later he would learn Chinese.
>
> His proficiency in language and culture was unique and in 1919 earned him an assignment as the U.S. Army's first language officer in China. In 1921, he volunteered to oversee an International Red Cross rural road-building project so he could interact directly with Chinese officials and laborers to hone his language skills and learn about their way of life. After his first year in China, Stilwell had become conversant in a notoriously difficult language and familiar with a culture that remained entirely alien to most Westerners.
>
> Stilwell spent most of the next 20 years in China, becoming one of the U.S. Government's most trusted China experts in the process. In 1926, he commanded a U.S. Army battalion near Beijing; then in 1935, became the American military attaché in China. After the attack on Pearl Harbor, U.S. Army Chief of Staff Marshall appointed Stilwell commander of the China-Burma-India theater and chief of staff to Chiang Kai-Shek, the leader of Chinese forces fighting the Japanese. Between 1942 and 1944, General Stillwell deftly used his knowledge of Chinese language and culture to build rapport with Chiang Kai-Shek, ensuring that the Chinese Nationalist forces remained a partner in the war against the Japanese.

4-30. There are many things that Army leader must be able to do, including—

- Operate in other countries and work within other cultures.
- Train, lead, and care for Soldiers.
- Exercise sound judgment and critical thinking to help accomplish missions.
- Develop effective plans.
- Manage and maintain equipment and other resources.

Self-development Goals

4-31. Self-development activities aim at learning new knowledge, gaining or enhancing skills, changing attitudes or values, or a combination of these. It is often easier to improve upon strengths rather than developmental needs. Learning is quicker and greater when strengths are used as a path to improvement rather than developmental needs. However, if a particular developmental need is an obstacle to development, consider improving it.

4-32. No set formula exists in choosing personal development goals. However, key considerations include—

- Personal strengths.
- Personal developmental needs.
- Family roles and responsibilities.
- Current or future roles.
- Army needs.
- Personal interests.

4-33. Personal experiences and goals, as well as personal interests, needs, and resources should influence the determination of self-development goals. Ideally, self-development goals will provide a long-term professional aim to work toward through a variety of activities. Figure 4-1 on page 4-8 provides an example of how to work through developing self-development goals.

1. Self-analysis

Strengths

The skill or ability that I am best at is—working with others who are very different from me.

The personal quality that I rely on most for my success is—flexibility and creativity.

I would love to learn more about—the Korean language and culture.

The activities I look forward to include—finding creative solutions to problems in the field, traveling, and meeting new people.

Others think the best job for me would be—Special Forces.

Weaknesses

The situation that causes me the most frustration is—doing paperwork and working in an office.

I am most hesitant when I try to—prepare the monthly law enforcement briefing.

Others think I am not very good at—administrative tasks.

2. Roles and Responsibilities

Soldier: Be as proficient as possible in my soldier tasks; serve the US to the best of my ability.

NCO: Lead and set the example for my Soldiers; Train and develop my Soldiers and myself.

Law enforcement: Respond to and resolve law variations; Be expert in the law and its application.

Active in community: Help plan and execute the church's food pantry; serve as assistant coach for youth baseball.

3. Army Need and Own Interests

Agile, adaptive leaders—Special operations

Special operations—Foreign languages and cultures

Total fitness, resilient Soldiers—Wilderness backpacking

4. My Developmental Focus

Learn more about the needs and opportunities in Army special operations, especially the role and use of language skills, cultural knowledge, and other regional expertise.

5. Set Developmental Goal

My self-development goal is to—prepare to meet the requirements for reclassification into MISO and become an expert on the Korean peninsula region.

Figure 4-1. Example of self-development goal development

MILESTONE PLANNING

4-34. After establishing self-development goals, create one or more milestones to get started and gauge progress. Use an IDP to document goals and milestones. Use each milestone to stretch you. Milestones can be a mix of short-term or long-term—whatever personally works and encourages progress. Milestones should—

- Be specific and measurable: They need to state what to accomplish so you can tell if you have met the milestone or not.
- Be meaningful: They should help achieve self-development goals.
- Provide a challenge: Milestones should stretch personal abilities and be challenging to accomplish. Challenging milestones increase motivation; being too easy or hard can hurt motivation.
- Have a time limit: Time limits provide motivation and will help gauge success.

- Be flexible: Build in some flexibility to overcome obstacles or revise milestones if necessary.
- Be realistic: Ensure milestones are reachable with available resources. For example, if a deployment will occur in the next 12 months, do not set a milestone requiring college attendance during that time. Keep in mind that unforeseen obstacles may occur along the way.
- Be cost effective: The benefits gained must be worth the effort, resources, risk, and other costs of reaching the milestone.

4-35. Every milestone requires at least a minimal amount of planning. After setting the first milestone, create a plan to achieve it. A plan can increase chances of success by—

- Identifying all required actions.
- Identifying the resources needed to meet the milestones.
- Establishing time estimates and deadlines that help track progress.
- Dividing large tasks into smaller parts to reduce being overwhelmed.
- Identifying possible obstacles and the actions and resources needed to overcome them.
- Making the best use of personal time and other resources.

PLAN TO MEET MILESTONES

Develop a plan by listing the first milestone and identify the main steps needed to reach it along with associated timelines to meet those milestones. Consider all of the developmental resources the Army has to offer as well as other sources to reach each milestone. Identify potential enablers and obstacles before beginning to better prepare for difficulties along the way. Collaborating through online forums and interest groups may help personal development and provide encouragement.

SELF-ENHANCED LEARNING

4-36. Self-development requires learning. Knowing how to learn is the most important skill required for self-development. Self-understanding, setting self-development goals, and planning milestones all influence a personal ability to learn. Beyond that, effective learning requires—

- Motivation and persistence.
- Learning opportunities.
- Effective learning methods.
- Deep processing.
- Learning through focused reading and analysis.

MOTIVATION AND PERSISTENCE

4-37. Self-development may require hard work over a long period, especially if the goal is to become an expert in an area or undergo significant personal growth. It takes motivation and effort to keep self-development efforts alive. Genuine motivation provides lasting energy because it is the internalization of goals and the desire to achieve them. Use these tips to stay motivated and to persist in the effort required to make significant changes:

- Recognize the benefits of self-development efforts. Think about—
  - Why the results are personally important.
  - How you will feel after reaching these self-development milestones.
  - The positive effect these efforts will have on others.
- Plan learning activities so that they—
  - Connect to the real world.
  - Teach knowledge, skills, or abilities that will help personal understanding.
  - Satisfy curiosity.

- Set specific and challenging milestones to progress through a personal developmental path.
- Milestones should—
  - Stretch enough to provide a sense of accomplishment and satisfaction after achieving them.
  - Not be so difficult to have a high chance of failing (know personal limits).
- Reward yourself for accomplishing learning tasks and milestones; it may be as simple as self-acknowledgement of completing a step.
  - Decide on the reward before beginning learning.
  - Keep the reward appropriate for the task.
  - Do not give the reward if planned tasks are not accomplished.
- Treat self-development as a duty. Make it routine—select specific times to work on self-development tasks.
- Maintain momentum.
  - Do not start a learning task then put it down for too long.
  - Work on the task a little every day until it is accomplished.
  - Break a big task into smaller ones to accomplish in a reasonable amount of time.
- Get support.
  - Find family members, friends, or supervisors for encouragement, accountability, recognizing accomplishments, and as a source of feedback.
  - Observe others who have successfully achieved their goals. Learn and model what they do.
- Review what has been learned so far.
  - Think about the progress made, personal growth, and resolved challenges.
  - Learn from mistakes and do not repeat them.

Setting Habits for Self-development

From a warrant officer:
Maintaining expertise and enhancing it are especially relevant to warrant officers who bring specialized knowledge and experience to their units. A practice that has served me well throughout my career has been to dedicate time each day for self-study. Goal setting was also important; it kept me focused and excited to move forward to the next challenge. On alternating days, I review relevant topics related to my area of expertise or general Army knowledge. I selected topics based on an honest assessment of my own strengths and needs, while mentors and raters recommended others. This practice began when I was a junior warrant officer, during field training exercises, deployments, and combat operations. As a senior warrant and senior leader within my organization. I still make time each day to learn something new.

Learning Opportunities

4-38. Learning stems from deliberate planned activities or from the unplanned experiences of daily life. Make the most of each learning opportunity, whether planned or not.

Teachable Moments

During a training center rotation, unit staff and subordinates assembled to conduct a rehearsal for an upcoming attack. By some accounts, the commander had been struggling for several days to keep up with the pace of operations. He opened the rehearsal by giving the briefest of intents then moved behind the sergeant major and crossed his arms. The brigade commander took note and listened to the intelligence brief. During a pause to reset the rehearsal, the brigade commander asked the battalion commander to join him at his vehicle. He asked if the battalion commander agreed with the assessment and then got to the teachable moment—'how do you, battalion commander, assess whether someone is on their game, whether they are engaged in and supportive of the plan?' The battalion commander paused and answered that it depends on whether they are actively listening, affirming what they agree with, and asking questions about the rest. The brigade commander replied that even the act of listening by a leader is a powerful motivator; that it can make the difference whether staff and subordinates stay focused and on task. He pointed out that their roles as commanders must be to set the example, especially when they are tired and in front of those who directly support them.

4-39. Leaders can embed planned learning into routine duties by using normal events as learning opportunities or it can be a completely separate, scheduled activity for a specific item. Prepare for the unexpected times by having appropriate learning materials available. It is a good idea to take advantage of time that opens up such as from transportation delays, waiting for appointments, or cancellations.

4-40. Unplanned learning happens when something unexpectedly captures your attention. Interest in the topic causes you to pay attention and learn. Attune your mind to draw attention to information related to self-development aims by thinking about developmental aims in detail—what you are trying to accomplish and why you want to accomplish these things. Review what you know and what you need to learn. Remind yourself of key terms and ideas related to the subject as well as who the experts in the field are.

PERSONAL AFTER ACTION REVIEW (AAR)

A personal AAR is an in-depth self-assessment of how leadership contributes to task and unit performance. Leaders should conduct their own personal AARs after a task is complete, or even while it is playing out, by asking themselves:

What happened and what were the consequences?

How were my leader actions supposed to influence the situation?

What were the direct results or consequences of my leader actions?

How did my actions benefit or hinder mission accomplishment?

How should I change my leader actions for better results next time?

What did I learn?

A good time to encourage personal AARs is following the unit AAR process. The unit AAR will clarify for the leader what happened and accurately assess mission accomplishment. Commanders can reinforce personal AARs by:

Walking less experienced leaders through the personal AAR.

Asking individuals what they learned from their personal AARs.

Telling subordinates the outcome of their personal AARs.

EFFECTIVE LEARNING METHODS

4-41. The purpose of each learning activity will help determine the learning principles to use to make the most of learning. The purpose may be to learn new knowledge, a new skill, or a new attitude about something. No matter the purpose, there are general principles of learning that apply:

- Use multiple senses. Memory stores information according to the senses (how it looks, sounds, feels, tastes, or smells). Moore senses used while learning enable better memory and information recall. Involve multiple senses by taking notes, highlighting, reciting, and observing.
- Space out learning sessions. Do not try to learn a large amount of information or a complex skill in one long session—try to break the material into multiple, manageable sections.
- Study the information or practice the skill on multiple occasions.
- Know the time of day when you learn best and study the most difficult material during that time.
- Design learning activities so that they mimic reality as much as possible. If the expected conditions to use the information cannot be duplicated, try to imagine the conditions as vividly as possible.
- Familiarize information through self-study prior to formal instruction. Reinforce learning by reviewing the information after instruction. This is a good way to review and test memory skills.
- When learning an entirely new field, go slow at first to ensure thorough understanding of the basics—it is important to have a solid foundation to build on.
- Learn in layers. Start with what you know to determine what is the first level of understanding, information, or skill needed. Learn that level then determine what the information just learned suggests to learn next. Each level builds on the previous and usually becomes increasingly detailed and interconnected.
- Learn like a scientist. Scientists adopt the attitude that the best knowledge is subject to change and that new discoveries may prove old beliefs or assumptions wrong. Start the inquiry with a problem

or question. Find evidence that answers the question and test possible explanations to gather evidence. It is also important to discover information that questions or refutes the possible explanations to avoid confirmation bias. Analyze the evidence and develop an explanation.

Principles for Specific Types of Learning

4-42. While the general learning principles apply to all types of learning, some learning principles apply based on whether the learning activity focuses on learning a new skill, a new attitude, or new knowledge.

4-43. Learning or improving a skill requires repeated, deliberate practice. Deliberate practice is not just repetition of a skill. Deliberate practice involves—

- Making your best attempt at performing the skill.
- Analyzing the results of the attempt (sometimes with the help of a coach or instructor) to identify ways of improving personal performance.
- Attempting the skill again using the identified improvements.

4-44. Learning a new attitude about something requires repeated exposure to and testing of the attitude. Taking on a new attitude might involve realizing that a prior viewpoint is counterproductive to obtaining goals. Changing attitude can be done in two ways:

- Behave as if you have already adopted the new attitude. If done often with positive results, it is likely that you will actually adopt the attitude.
- Observe another person behaving in a way that reflects the attitude. If you respect this person as a role model and you see the person gaining some benefit from the behavior, you may eventually come to accept and adopt the attitude for yourself.

4-45. Learning new knowledge requires linking the new information to already known information. This occurs by deeply processing the information that you want to learn. Deep processing and the related mental skills of critical and reflective thinking are detailed in the following section.

DEEP PROCESSING

4-46. The ability to learn and recall information depends upon what someone does with the information while trying to learn it. Deep processing requires analyzing the new information, picking it apart, using it, and connecting it to already-known information. There are many ways to practice deep processing:

- Relate the new information to known information. This is the most important factor in learning. An increased number of connections create multiple ways to recall the new information.
- Think about conflicts between the new information and any prior understanding of the topic. Resolve the conflict in your mind and be able to explain the conclusion and resolution process.
- Summarize the material you are learning in your own words.
- Study the structure of the subject to see how facts, ideas, and principles relate to each other. Draw pictures or diagrams that show the connections between components. Diagrams can also be useful for learning the steps in a process.
- Organize new information into categories. For example, group illnesses by their symptoms, exercises by their physical benefits, or aircraft by their purpose.
- Ask and answer questions to make new facts, ideas, and principles useful and important:
    - How does this relate to prior knowledge?
    - What does this imply?
    - What other examples of this can I remember?
    - Why is this important to me (or others)?
    - Where else could this apply?
    - Where or how could this be used?
- Think of what the new information reminds you of. Develop metaphors and comparisons.
- Explain or teach the material to another. Did you get it right? Was it all covered? Does your explanation demonstrate personal understanding? Using the new information with others will test your mastery of it, and their reaction may help you better understand the material.

- Argue one or both sides of an issue. Do this to help you think about the truthfulness of a position. Get your ego involved by imagining you are in a hotly contested debate or trial.
- Personalize the information by relating it to experiences or future expectations.

4-47. Using critical and reflective thinking skills is essential to being an effective learner and gaining subject expertise. Critical thinking and reflective thinking do not apply solely to learning but are essential practices and important ways of deeply processing information for duties across the range of military operations.

Critical Thinking

4-48. Critical thinking involves questioning what is seen, heard, read, or experienced. Critical thinking ensures that the person is engaged in the learning process, critically considering the information or practice of skills. Critical thinking requires analysis, comparisons, contrasting ideas, making inferences and predictions, evaluating the strength of evidence, and drawing conclusions. It also requires the self-discipline to use reason and avoid impulsive conclusions. These questions can guide critical thinking—

- What is the evidence for and against this conclusion?
- What are the alternative or competing theories, explanations, or perspectives?
- Why is this important?
- What are the implications of this?
- Is the logic of the argument or reasoning sound?
- Do the numbers, quantities, and calculations make sense?
- Do the supporting facts agree with other sources?
- Why or how does this work?
- How likely is this?

Reflective Thinking

4-49. Closely related to critical thinking, reflective thinking seeks to build understanding, interpret experiences, and resolve questions. Reflective thinking requires thinking through the gathered information in detail to organize it, apply principles, make connections, and form conclusions. Use these questions to organize personal thoughts—

- What does this information mean?
- What conclusions can be drawn from this?
- How can this information be used?
- How does this fit with my existing knowledge and experiences?
- What are the implications of this for others or me?
- What is the big picture and how does this fit into it?
- What is the best way to learn about this subject?
- Where should this take me in my studies and self-development?

Learning through Focused Reading and Analysis

4-50. Books and other written materials may be key learning resources for self-development. To maximize learning, approach reading for learning differently than casual reading. Deep processing of written materials is essential to the ability to understand, recall, and use the information contained in the books and other documents. Even though books may present information in a logical way, you must take an active role in teaching the information to yourself.

4-51. The Survey-Question-Read-Recite-Review method uses the deep processing principles. Developed over 70 years ago, these activities comprise one of the most widely recommended and effective ways of learning from written materials.

4-52. *Survey.* Before beginning to read, look over the chapter, article, or other material to build a mental framework or outline of the material and establish a purpose for reading it. This mental framework will help understanding the purpose of the material, set expectations so that attention will be drawn to important information, activate memory of what is already known, and give a skeleton of understanding to add to while reading. Survey the material by leafing through it and doing the following—

- Make predictions about what the sections of the document will discuss. Complete accuracy is not necessary, but it helps active understanding of the material.
- Note the title, headings, and subheadings to see the sequence of topics and their relationships.
- Look at graphs, charts, diagrams, and pictures and read their captions.
- Read quotations, vignettes, and other short statements that are set off from the main text.
- Scan footnotes to get a sense of where ideas come from or what they mean.
- Note emphasized words and phrases (such as bold, italic, underlined, or colored text).
- Read the introduction, abstract, and summary if available; if not, read the first and last paragraphs.
- Read the first and last sentences of each paragraph.
- Review other learning aids that the material may have, such as study guides, advance organizers, chapter outlines, learning objectives, or review questions.
- Decide what you want to learn from the material.

4-53. *Question.* While surveying the material, write down questions that you want to answer while reading the material. Developing questions to guide your study increases interest in the material, makes you alert to important information, helps you stay involved with the material, and relates the new knowledge to what you already know. To develop questions—

- Turn the title, headings, or subheadings into questions. For example, if a subheading is "The Four-Step Calibration Process," a question may be, "What are the four steps of the calibration process?"
- Ask questions about graphs, charts, diagrams, and pictures. For example, a graph showing an increasing rate of traffic fatalities in the United States could lead to the question, "Why have traffic fatalities increased in the United States?"
- Consider questions that the author includes in the document, such as in call-out boxes or review questions at the end of a chapter. It can help to rephrase these questions so that they are meaningful and easier to remember.

4-54. *Read.* Read the material one section at a time. Use multiple senses by reading, taking notes, highlighting, and maybe even reading aloud. These tips will aid understanding, retain interest, and retain the information:

- Look for the answers to your questions and write them in your own words.
- Look for additional questions to answer and important information that you had not anticipated.
- Use deep processing to relate the new information to things you already know.
- Highlight important information, especially information that answers questions you wrote.
- Write notes in the margins of the document or on separate paper. These can be key words or phrases, definitions, reminders to guide studying, and other useful points to remember.
- Make diagram that show how a process works, timeline, sequence of events or the relationships that exist between different components.
- Respond to points made in the document by noting ideas about them in the margins. This will help personalize the information and relate it to information already known. For example, notes may highlight disagreements with a point, how a stated idea relates to another idea learned elsewhere, gaps or questions that remain in the information, or implications of the information.
- Look for connections, discrepancies, comparisons, and relationships between information presented in the document and other readings, lectures, or personal experiences.

4-55. *Recite.* Reciting tests knowledge and understanding of the information. Self-testing is a method of deep processing that can enhance memory. Reciting helps ensure minimizing knowledge gaps.

- Stop reading at the end of each section and summarize the material in the section from memory.
- Ask yourself the questions you previously wrote for that section.
- Explain charts, graphs, diagrams without referring to the text or personal notes.
- If you have problems, go back and review the section until you can recite its important information and concepts from memory.

4-56. *Review.* Reviewing helps refresh and strengthen memory and mastery of the material.

- Review immediately after reading the entire article or chapter. Review the document again in 24 hours and again several days later.
- After reading the entire article or chapter, flip back through it, looking at headings, subheadings, graphs, charts, diagrams, and so on. See if you can recall the important information for each item. Study the material to fill in any gaps.
- Go back through all of the written questions and see if you can answer them from memory. Study the material to answer any missed questions.
- Explain how all of the sections fit together. What are the overarching points and principles?
- Explain how the information in this document relates to self-development goals.
- Interaction with a friend who has studied the same information can help maintain focus, provide different perspectives on the material, and clarify difficult or misunderstood points.

Personal Reading

4-57. Documents often suggest related information to expand knowledge of the subject. The end of a chapter or book may list related documents. The bibliography or footnotes identify information sources the author used. To help narrow the search, make notes of any reference that sounds interesting and relevant.

Reflective Journaling

4-58. A journal may track and record the occurrence, actions, and outcomes of various situations. Reflective journaling goes beyond a personal AAR including periodic entries on self-awareness of personal strengths, developmental needs, values, feelings and perceptions, and questions and ideas about leadership situations. A leader may track personal successes and lessons learned by recording their experiences in leading others, the chosen actions, the resulting outcomes, and any insights. The journal may serve as a reference to pass along lessons learned to others. Key leader references also may be recorded.

4-59. Sample reflective journaling questions include—

- What is the best thing that happened today or this week?
- What is the most difficult or satisfying part of my work? Why?
- What do I think is my most valuable or valued contribution?
- What compliments and criticisms have I received lately? What did I learn from them?
- What did I learn because of a recent disappointment or failure?
- How do recent experiences connect to my long-term goals?
- What risks have I taken (or avoided taking) lately?

4-60. Individual leaders should decide whether to share their journal content with their immediate leader or others. Leaders can reinforce reflective journaling by—

- Carrying a journal and being seen writing.
- Citing lessons learned while referring to journal entries.
- Providing time for a leader to reflect and write down personal lessons learned.
- Providing leaders with a journal and a personal note encouraging them to use it.
- Asking leaders to write or recount a story of a key leader challenge and use the stories to pass on lessons learned to less experienced leaders.

LEARNING IN ACTION

4-61. Self-development efforts take time and effort. To stay on track—

- Let milestones serve as a guide.
- Overcome self-development obstacles.
- Work efficiently.
- Maintain forward momentum.
- Assess progress.

- Make course corrections.
- Set the next milestone.

LET MILESTONES GUIDE

4-62. Use the milestones as a guide to—

- Avoid impulsive actions that may be ineffective and discouraging.
- Keep the big picture in mind.
- Work effectively toward self-development goals.
- Take advantage of resources and overcome obstacles.
- Measure success.

4-63. Adjust the plan as needed to reach milestones. Be willing to update the plan to improve it, change goals, address obstacles, take advantage of resources, and reflect upon accomplishments.

SELF-DEVELOPMENT OBSTACLES

4-64. In developing a milestone plan, obstacles to reaching the first milestone were identified. There is always the possibility of encountering internal and external obstacles, despite thorough preparation.

Internal Obstacles

4-65. Procrastination, apathy, and pride are major obstacles to self-development and occur for many reasons. Some come to realize their milestones are too ambitious, complex, unclear, or difficult. Others hesitate because of the effort or discomfort that the work requires or lack the motivation to start. These techniques address procrastination:

- Write it down:
  - Write down the goals and milestones and post them where they will be seen frequently.
  - List the benefits of doing the work.
  - Write down the next planned action and associated deadlines.
- Involve others:
  - Tell others about personal intentions and deadlines.
  - Talk through the task with someone else.
  - Schedule time with someone else to study or work together.
- Break it down:
  - Break big tasks into smaller, manageable tasks.
  - Make a list of the small steps required to accomplish each milestone.
  - Start with easy steps then gradually build to steps that are more difficult.
  - Mentally rehearse a difficult task or talk through the task with someone else.
- Make a routine:
  - Pick a routine time to work on self-development activities.
  - Use good time management skills by following a dedicated schedule.
  - Plunge into the task immediately to gain momentum, keep it going.
- Know yourself:
  - Know your habits. Recognize what you do to avoid things you do not want to do.
  - Confront yourself when you see yourself doing these things.
  - Identify self-defeating attitudes and replace them with positive ones.
  - If a task is repeatedly delayed, do you really intend to do it? If not, remove it from the plan.
- Be open to deviations in plans and milestones. If an area is overly complex or not interesting, then consider an adjustment for a higher potential path.
- Get motivated.

4-66. A poor attitude also can interfere with learning and make it difficult to understand and remember information. For example, thinking that math is hard or disliking history can interfere with an ability to learn anything related to math or history. Other attitudes, such as closed mindedness, inflexibility, or rigid adherence to beliefs and assumptions, can interfere with learning. To combat poor attitudes, identify a productive replacement. Practice thinking and behaving with a positive attitude until it feels natural and becomes a habit.

External Obstacles

4-67. External factors such as workload or other personal or professional obligations may hinder self-development efforts. Resistance may also come from others, such as a spouse who resents time spent away from the family or friends who may pressure you to spend time with them.

4-68. A lack of resources is another common roadblock. Resources include anything needed for self-development including people (such as teachers, coaches, and mentors), facilities (such as schools, libraries, and museums), and things (such as training programs, books, and equipment). Learners best handle external obstacles through careful planning and creativity.

Work Efficiently

4-69. By efficiently managing workload and personal life, one can increase how much time is available to spend on self-development. To increase efficiency—

- Take care of yourself. Proper food, exercise, and rest enable functioning at your best.
- Manage time and energy efficiently. Keep a running 'to do' list. Prioritize each task according to its importance, required work, and completion date. Remove low-priority tasks from the list.
- Look for ways to accomplish daily activities and routines in less time. For example, combine several errands in a single trip instead of making individual trips.
- Learn to quickly locate and obtain the information needed for self-development and other requirements of daily life.
- Organize work and living areas so that required information, tools, and workspace are available.

Forward Momentum

4-70. It is important to keep the developmental momentum moving forward. There may be a tendency to slow down after completing an important self-development step or be discouraged by setbacks. Resting after a strong effort is natural, but too much rest may make it hard to restart. Maintain momentum by—

- Keeping a positive attitude: Let go of setbacks and start each day with renewed enthusiasm. Each morning offers an opportunity for a fresh start.
- Making consistent progress: Try to accomplish something, however small, related to self-development milestones and goals each day.
- Recognizing benefits: Benefits can be tangible results such as increased pay, awards, and abilities or intangible results such as pride, a sense of accomplishment, and satisfaction. Remember that important benefits often require hard work.

Progress Assessment

4-71. Assessing progress can provide encouragement to keep going if things are going well or to guide changes if they are not. Individuals can assess progress at any time—while working toward a milestone or after completing one. To assess progress—

- Use objective and subjective measures.
  - Objective measures are things that can be seen or expressed in numbers, such as a test score, time required to perform a task, number of books read, or number of college credits earned.
  - Subjective measures are things that cannot be easily observed or expressed in numbers, including feelings of satisfaction, accomplishment, personal growth, or difficulty. Subjective

assessments of progress can come from both personal judgment and from feedback. Subjective indicators are often sufficient to track most self-development activities.

- Compare the milestone plan with what actually happened and adjust the remainder of this milestone plan or future milestones to account for lessons learned.
  - *Timeline*: Was the timeline met? If well under or over the timeline, determine why. Maybe the timeline was not reasonable, more or less work was anticipated, received extra help, encountered obstacles, or the material was more involved than initially thought.
  - *Action Steps*: How successful were you in accomplishing the steps identified for reaching the milestone? What helped or hurt success? Were the identified steps the right ones?
  - *Resources*: Were the types of resources needed to achieve the milestone correctly identified? Did the plan omit any resources? Were necessary resources available? Are there any other resources that might have worked better?
  - *Obstacles*: Were identified obstacles encountered and was the plan for overcoming these obstacles successful? Were unexpected obstacles encountered?
- Decide if you are satisfied with your progress or if the milestones or general self-development goals need changes. Indicators to consider in making a course correction include:
  - Unsatisfactory progress.
  - Too much stress or effort required to complete developmental activities.
  - Loss of interest in achieving self-development aims or change in the benefits expected from achieving those aims.
  - Changes in professional or personal situations that conflict with self-development activities.
  - Being dissatisfied with personal development.

COURSE CORRECTIONS

4-72. Self-development occurs over time in a dynamic environment that includes professional and personal responsibilities. At some point, obstacles or other challenges will force a change of plans. If the progress assessment indicates course corrections are needed, determine what correction is warranted:

- Goal: A self-development goal or milestone may have turned out to be too difficult, too easy, or just not what was hoped. Examine other possible self-development goals or milestones. Identify why the unsatisfactory goal or milestone was selected and avoid repeating any missteps.
- Action Steps: If the actions taken to achieve milestones were not effective, figure out why they did not work, and then develop actions that are more effective. To be effective, you must be capable and willing to perform the actions with available resources. If a course correction is required due to obstacles then create new action steps that avoid or solve these obstacles. Action steps should form a logical path from where you currently are to achievement of the milestone.
- Resources: The identified milestone resources may have been inappropriate, inadequate, or unavailable. If so, analyze planned action steps to determine the resources (such as time, money, equipment, facilities, or help) needed to perform these steps. Determine if they can be obtained.

THE NEXT MILESTONE

4-73. With the first self-development milestone achieved, a full cycle of self-development is completed. It is now time to continue the self-development process by setting and pursuing the next milestone.

This page intentionally left blank.

Chapter 5
Unique Aspects for Development

5-1. Character, judgment and problem solving, and adaptability are capabilities that are especially valuable to leaders and team members in special situations. They allow leaders and teams to address the demands of complex, ambiguous, and chaotic environments of military operations. Whether making the tough moral decision, thinking critically to resolve uncertainty, thinking from a broad and strategic perspective, or adapting to unexpected changes, expert leaders find the way to do what is right. This chapter describes these capabilities and identifies unique aspects of developing, enhancing, or fostering them in leaders and teams.

CHARACTER

5-2. Character is a critical component of being a successful Army leader. Character is one's true nature including identity, sense of purpose, values, virtues, morals, and conscience. Character is reflected in an Army professional's dedication and adherence to the Army Ethic and the Army Values. Character is the essence of who an individual is, what an individual values and believes, and how they behave. Doing the right thing the right way for the right reasons demonstrates character. Demonstrating character often means resisting the easier wrong in favor of the tougher right. Making the right choices involves discipline. Discipline can be thought of as the foundation of character. Team character is the melding of individuals' character in a team.

5-3. As the uncertainty of operating environments dictate, junior leaders need to be capable of independent decisions using sound discretionary judgments founded in moral character. Character is also such an important quality of a leader because decisions and actions of the leader are viewed by others. The demonstrated character of the leader greatly influences how other people either emulate their conduct or disapprove of it. These can, in turn, add to or detract from team trust and cohesion.

5-4. Character forms over time through education, training, and experience in a continuous, iterative process. Leaders employ character when all decisions, big or small, are analyzed for ethical consequences. One must have the knowledge of how to address the consequences. This knowledge comes from the Army Ethic, personal experience, and others' guidance. Internalizing the moral principles of the Army Ethic as personal beliefs is essential for character development. An individual demonstrates character when they correctly identify the ethical implications of a decision, possesses knowledge to act, and acts accordingly.

5-5. Leaders enable the development of character in others by conveying clear ethical expectations, modeling right conduct, and establishing discipline to uphold the Army Ethic and embody the Army Values. This comprises the essence of what it means to be an ethical leader and serve as a powerful influence on character development. Guided by the Army Values, character development is founded on discipline. To develop character in others, leaders must embody the Army Values every day. Leaders must communicate expectations that others embody the Army Values as well. To reinforce desired behavior, leaders should recognize Soldiers who demonstrate exemplary conduct. When a Soldier falls short of the Army Ethic or the Army Values, leaders must counsel them and help them see the gaps between their personal values and those the Army espouses. Leaders look for the character developmental opportunities that exist in day-to-day operations. In this way, Soldiers learn what is expected of them. Reinforcing ethical standards increases the likelihood of ethical decisions and actions, and promotes an ethical climate.

5-6. Leaders shape the ethical climate of their organization while developing the trust and relationships that enable proper leadership. Over time, the fostered ethical climate contributes to enhanced organizational ethical behavior. The internalization of ethical principles develops as the culture reinforces the acceptance and demonstration of ethical behavior. All Soldiers contribute to their team's character. Modeling positive character fosters social norms and expectations to choose the tougher right over the easier wrong. Team members hold each other accountable to embody the Army Values and demonstrate character at all times. This approach to character at the team level strengthens team trust and cohesion.

JUDGMENT AND PROBLEM SOLVING

5-7. Intellect enables a leader to understand, visualize, and decide and is essential in unfamiliar and chaotic settings. Judgment, as a key component of intellect, is an ability to make considered decisions and come to sensible conclusions. Leaders can reflect on how they think and better foster the development of judgment in others. Problem solving, critical and creative thinking, and ethical reasoning are the thought processes involved in understanding, visualizing, and directing. Problem solving involves situation assessment (understanding), imagining (visualizing), and converging on a solution (directing). Thinking critically involves analytical, cautious, and convergent judgment. It checks on the sensibility, relevance, and relationship of meaning and possibility. Creative thinking is generative, daring, and divergent. Critical thinking considers what might be wrong, while creative thinking considers what is possible. The two complementary processes—evaluation and generation—occur in a free-flowing manner depending on what ideas and conclusions stem from thinking in specific situations.

5-8. A goal of all leaders and teams is to think as well and as thoroughly as time permits. The brain encodes experience as expertise that allows automatic and intuitive responses, which frees up time to apply to other thinking or provides a reserve capacity for addressing the most complex problems. Intuition can operate rapidly, but the downside is that it can be misapplied. Intuition operates based on the best or closest match, having no built-in or automatic process that checks on mismatches in cues, and no repair of ill-fitting ideas. Evaluation, repair, and design are roles of deliberate thinking processes.

5-9. Leaders draw on their knowledge and expertise in the context of each part of a problem. However, most situations will always have incomplete knowledge. Thinking is a technique to identify gaps in knowledge. Experience or a hunch can be used to facilitate a new way of framing (seeing or structuring) a problem or a solution. Leaders test ideas through visualization or a war-gaming process. The thought process judges how well ideas meet goals. (See emerging doctrine on the Army Design Methodology.)

5-10. Everyday thinking switches back and forth from a subconscious process of intuition to deliberate, effortful thought. The active monitoring of one's own thinking guides the process, keeping it on track. Thinking about thinking is metacognition. Metacognition and deliberate thought are processes that people can learn to improve. Intuition develops through the natural accumulation of experiences. Individuals develop judgment intentionally through overt attention to the deliberate side of thinking. The following sections describe these deliberate processes.

THINKING ABOUT THINKING

5-11. Thinking about thinking is one way to develop better judgment. Metacognition involves both self-awareness and self-regulation of thought. Metacognition is important to military leaders dealing with complex problems because it involves adapting to the situation. By increasing the awareness of one's own thinking, mental capabilities can be allocated to the pressing problems at hand. Being self-aware means having insight into how one learns, and the thought patterns and strategies that are typically used when thinking. Being better in touch with how one thinks can increase the chances for successful thinking. To improve thinking capacity for good judgment and to self-regulate thinking in the moment, leaders should practice thinking about how to solve problems and how to decide.

> Memory and thought processes are complex, but consider if they were simply files. An increased number of files become available as the individual studies and learns. The more often the individual accesses the files, their contents become more familiar, and chances increase that a file will be the best match to a future situation.

5-12. Improving judgment requires self-reflection and hard work to adopt new habits. Making thinking more deliberate will prompt self-reflection. Through practice, new ways of thinking will become easier to use in daily operations and especially in pressure situations where they are most beneficial. Improved thinking strategies will create greater self-confidence, making it more likely to address rather than avoid complex challenges. Table 5-1 provides questions to help leaders reflect on their thinking and develop better judgment.

Table 5-1. Examples of reflective questions

For improved understanding ask:	What is this situation? What other situation is like this one? What is this situation not like? What do I know about situations like this? How could this situation happen? How should I think about this situation to define the problem or opportunity? What is the real problem? What do I not know that I should?
For improved visualization ask:	What else could this situation or solution be? Are there any assumptions unneeded, new ones needed? What constraints are there? What needs to be accomplished? What is likely to happen? How should I prepare for future situations?
For improved decisionmaking ask:	What is the solution or plan? Does a solution dominate others? Can the solution be redesigned so that it does? Is there a specific way to reason and decide about the solution? What would my enemy not want me to do?

Critical Thinking

5-13. Critical thinking is composed of various techniques to consider the soundness and relevance of ideas as they apply to understanding a situation or determining a way ahead. Teams that engage in critical thinking make assumptions explicit and identify differences and similarities in how facts apply to the situation. Critical thinking is an active process in situation assessment that seeks to obtain the most thorough and accurate understanding possible. Situation assessment is a dynamic process that requires time and effort. Practice develops skill at critical thinking. Skill will facilitate the ease and smoothness of application to assessment and problem solving. (See ATP 2-33.4 *Intelligence Analysis* for information on critical thinking processes.)

5-14. High performance teams demonstrate mental agility (see ADRP 6-22) in their willingness to approach problems from different viewpoints and to hold and work on opposing ideas until identifying the best solution. High performing teams adopt the practice of using different perspectives in their critical thinking. Leaders can encourage critical thinking by how they challenge and pose questions to their teams. The leaders best at developing others actively lead the team to consider alternative points of view, multiple contingencies and first, second and third consequences of multiple courses of action. Teams that practice critical thinking and reflect on it will broaden their capabilities for tackling complex problems—difficult to solve because of incomplete, contradictory, or changing requirements.

Creative Thinking

5-15. Creativity is largely an attitude. To become more creative, leaders must be willing to make unusual connections that defy convention. They must be prepared to accept the risks of being different or wrong. Unusual connections may arise out of either effortful thought or from a relaxed, open state. Creative thinking involves examining problems from a fresh perspective to develop innovative solutions. Creative thinking occurs by consciously generating new ideas, and re-evaluating or combining old ideas, to solve a problem. Creativity is a willingness to accept change and apply a flexible outlook for new ideas and possibilities.

5-16. Looking at problems from different perspectives can improve one's understanding of a situation. It can lead one to see new goals and available options. Choosing to take multiple perspectives helps to understand situations, find new or creative solutions, and evaluate solutions. Any shortcoming or restriction in one's perspective is a possible source of problems in reasoning. Problem solvers can adopt different perspectives by taking on the role of another (such as the enemy, a neutral bystander, or adjacent unit commander), using new or different frames of reference, shifting importance about various problem elements, or reversing the goal. These require an openness of mind willing to apply a different perspective and practice in shifting perspectives. Adopting different perspectives is a way to enhance creativity and critical thinking.

5-17. Identifying hidden assumptions can be useful for developing greater creativity and insight. Coming up with reasons against a preferred conclusion or option instead of in favor of that conclusion or option will improve how thorough reasoning is done. This will also help identify contingencies that may occur. One can force oneself to imagine what causes a speculative conclusion to be incorrect. Considering ways something would not be true, allows determination of other possible aspects of a situation and ways to shape the outcome to avoid those undesired states.

Finding hidden assumptions or imagining failure are similar techniques that protect against group think and hasty agreement with conventional wisdom. To check for hidden assumptions, start with an assessment or course of action, consider that it is not true or has failed. Force yourself to think about what caused it to *fail*. Those causes are likely to be assumptions that were not evident.

Finding Hidden Assumptions

You are in the defense overlooking a river. The intelligence analysis has the staff convinced that the enemy force now advancing toward the river will pause on the far side while his forces close there massing combat power before launching the attack across the river.

Caution! The more confident and unanimous you are in a prediction, the more vulnerable you become to the consequences of error. Check for hidden assumptions used in arriving at the conclusion that you were not aware you were making. To do this, suppose for a moment that the conclusion is wrong, and ask, "under what conditions might the enemy NOT pause, close, and launch the attack?" Maybe it is a feint. Maybe we are responding to their approach exactly as they hoped we would. We assumed they were not sophisticated enough for such a scheme of maneuver, but how much do we really know? The assumptions that we did not realize we were making could prove incorrect. Their commander could be an expert and audacious tactician. Identifying hidden assumptions can mitigate the risk of an incorrect assessment.

Critical and Creative Thinking Development

5-18. Critical and creative thinking come together as practical thinking that captures the strengths of how people approach everyday problems, calling on experience over formal models such as classical logic. Creative thinking techniques help generate new insights. Critical thinking brings out differences that are not normally obvious. Both types of thinking fill in gaps in knowledge and resolve uncertainty. Signs of a practical thinker include a willingness to try alternate approaches to thinking, being open to others' positions, being prepared to think about issues instead of ignoring or dismissing them, and asking insightful questions.

5-19. Leaders should develop critical and creative thinking in team members. These abilities and capacities for intellectual and critical thought are essential to effective problem solving. The actions of subordinates, based on their own critical thinking skills, will often dictate the success of the team. One of the best ways to develop critical thinking in an organization is through example, by being a critical thinker. Thinking critically and setting conditions that encourage others to think critically are effective ways to enhance the process by team members. Leaders should be willing to take risks and encourage prudent risk taking by others. Thinking critically and creatively and thinking about thinking can cause people to question their own abilities. Leaders can counteract the unsettled feeling by listening attentively, affirming their subordinates' abilities, and reflecting about the processes of thinking and successful outcomes achieved from thorough thinking.

5-20. How people think and feel about learning and knowing affects their critical and creative thinking and development of judgment. For example, an attitude that thinking can resolve problems will lead to better results in overcoming difficulties through thinking. Attitudes that conflict with sound thinking should be diminished. These attitudes include feeling that changing one's mind is a sign of weakness, that being open and deliberating among options leads to confusion, that quick decisionmaking is how one demonstrates expertise, and that truth comes from authority.

5-21. Positive attitudes that contribute to developing critical and creative thinking include—

- Persistence. If one line of thought or action is not working, then finding another line may work.
- Willingness to expend effort. A willingness to engage in deeper, more thorough thinking is important for critical thinking, even when the effort may not initially seem useful.
- Active fair-mindedness. Taking special effort to find out whether one's ideas will work by imagining what is wrong with them is a good way to be fair-minded. Using the same standards, regardless of the issue or who supports a position is another quality of fair-mindedness.
- Detachment of ego. Keeping reasoning separate from self-esteem helps guard against being caught up in being on the right side of an argument or rationalizing why failure was out of one's control.
- Tolerance of uncertainty. Believing it is fine not to know something is a positive characteristic. Yet, being motivated to resolve uncertainty once it is recognized is even more important. There is an advantage to having to think through problems to figure them out, instead of using minimal, surface cues that could lead to interpreting a situation incorrectly.
- Openness. Being open to different and multiple possibilities leads to better decisions.
- Retraction of commitment. Willing to change beliefs about a preferred solution or a problem viewpoint is an attitude that has positive effects.
- Flexibility of process. Realize that standard processes will not work for novel, ill-defined, or complex problems. Adapting or discovering a new way of thinking that will help reach a solution.
- Willingness to learn. It is natural for leaders to feel an expectation to have the knowledge and experience to perform well. Being willing to engage in learning is adaptive. One characteristic of experts is that they understand what they know and what they need to learn.

5-22. Thinking ahead and predicting potential ways that a situation assessment may be wrong or that a course of action could depart from the anticipated plan will make leaders better prepared to handle the unknown. Having identified and thought about various contingencies better prepares the team for what could occur.

AMBIGUOUS OR UNFAMILIAR SITUATIONS

5-23. Deliberate effort applied to thinking is a way to provide best guesses about ambiguity—uncertain situations, uncertain actions by an adversary, and uncertain outcomes. One way to be prepared is to have a standard set of questions to ask oneself when faced with uncertainty. Developing a practice of asking questions will prompt additional reasoning. Table 5-2 provides some example questions to focus thinking.

Table 5-2. Example questions to focus thinking

Question prompt	*Example*
What if…?	What if… this assessment were not the case?
What else…?	What else… could be happening?
So what?	So what if that happened… is there a meaningful difference?
What are the specifics?	Can claims be confirmed with specific information?
Is there a weak link?	Are there any inconsistencies or confusions?
What is unexpected?	Are there new conclusions or implications?

5-24. A natural tendency is to discount information when it does not fit into expectations. Some refer to this as a confirmation bias, but that reference implies a conscious or active process when it usually is not. It is difficult to undo individual's assessment or conclusion once it forms. The first step to protect against the discounting of unexpected signals is to pay attention to information that does not fit expectations. Assessment of the information can indicate whether an understanding needs to change. If no definite determination can be made, even heightened vigilance for changes should occur.

5-25. Dominance structuring is a useful way of thinking for designing a solution and helping ensure it is the best available solution. The dominance structuring technique can be used as part of a formal military decisionmaking process as the 'nuts and bolts' way of thinking and iterating through course of action development, analysis, comparison, and wargaming. The technique can be used in an accelerated mode as a mental thinking drill.

DOMINANCE STRUCTURING

First, consider the relevant dimensions of the problem.

Identify the initial most promising alternative solution by eliminating alternative solutions that are unattractive on important dimensions.

Choose an alternative if it is better than all others on at least one dimension and equal to other options on other dimensions. This will be the dominant solution.

If the most promising alternative does not initially dominate all others, then reconsider advantages and disadvantages relative to other possible solutions.

Modify the most promising alternative until it dominates other alternatives. This will be the dominant solution. If no dominant solution appears, reconsider what are the most important dimensions of the problem and repeat the dominance comparisons among alternatives.

STRATEGIC THINKING

5-26. Strategic thinking is an imperative for military leaders to shape the future of operations and to steward resources at their disposal. Strategic thinking is valuable in all levels of leadership. It is important to take time to think of the overall view and to make decisions that set the stage for plans lasting years. Strategic thinking generates a cohesive understanding of situational dynamics presenting options of advantage and long-term organizational success. Thinking skills and activities directed at outcomes that produce an overarching approach or plan to achieve a particular aim characterize strategic thinking. In this case *strategic* describes the type of thinking rather than the usage in joint doctrine to describe a level of war, security objectives, or ideas to employ the instruments of national power. In contrast to thinking following traditional problem-solving steps, strategic thinking is not looking to solve a bounded problem but is creating a set or pattern of decisions to achieve future success. While a tactic is a specific prescription of how something will be done, a strategy is a philosophy of what is valued and consists of guidelines or boundaries that shape what actions to take.

5-27. Clearly, strategic thinking is an important skill for senior leaders who must establish high-level goals and broadly scoped policies. However, strategic thinking is also important for lower level leaders when they address recurring problems and consider enduring, robust solutions. The earlier leaders are exposed to strategic thinking, the more likely they will try it, apply it, and, over time, get better at it.

5-28. Strategic thinking can be broken down into several activities:

- Situational understanding. Understanding is enabled by scanning the environment for recurring, novel, and key cues that are integrated and used in sensemaking, predicting, and testing what exists. Visualization is a related activity used to fill in gaps of knowledge about what exists or used to consider what will exist in the future. Subskills include discriminating among relevant cues, integrating diverse information, applying mental war-gaming, and modeling.
- Questioning. Asking questions demonstrates an openness to different perspectives and a desire to consider alternate or unconventional assessments. Questioning is also a key component of thinking critically by improving the thoroughness of judgment. Consistently demonstrating a willingness to shift perspective, to look for alternate explanations, and avoid mindsets and fixations characterize cognitive flexibility.
- Systems thinking. Systems thinking involves considering the factors of a situation or a solution as a system of interrelated parts with inputs, processes, outputs and feedback. A systems orientation operates from a view that an understanding or model can be created or used to explain—or improve

upon—what occurs (as applies to situational understanding) or what could occur (as applies to problem solving). See Army Design Methodology doctrine for more about systems thinking.

- Analogical reasoning. Thinking that deals with complex problems with unfamiliar or unknown conditions and outcomes occurs by drawing on current knowledge. Analogical reasoning is a specific approach where known or similar concepts and relationships map to what is yet not understood. Historical comparisons are useful in strategic thinking to consider what has occurred under one known set of conditions.
- Self-awareness. Since strategic thinking involves unknowns, multiple paths, trials of what might exist in a situation, and possible results of a solution, an ability to manage personal thought processes is important. Metacognition is being aware of what oneself is thinking, what one knows, progress toward a conclusion, and in testing strategic approaches and conclusions about them.

5-29. The development of strategic thinking occurs largely by addressing complex, dynamic challenges and practicing critical and creative thinking. One learns strategic thought by working in context and becoming skilled at basic aspects of situational understanding and visualization. Leaders or instructors can accelerate the development of subordinates' thinking by assigning projects or duties with opportunities for strategic thinking. For a master sergeant it may be developing training plans for a specific system or developing and assessing the long-term goals of a remedial fitness program. For junior captains it may be assigning them to analyze and develop a plan for managing range use. For Army Civilians in an initial administrative position it may be assigning them to an installation task force on energy reduction. Professional military education courses reinforce strategic thinking by assigning projects requiring the application of the skills and grading how well a student employs them. For example, the Army War College started with students solving real-world problems as an extension of the War Department and now employs similar methods.

ADAPTABILITY

5-30. A key outcome of development of an individual leader or unit is building increased capability to adapt to meet mission challenges. Adaptability for the purpose of performance is an effective change in behavior in response to an altered or unexpected situation. The Army stresses the importance of adaptability due to the rapid pace of world events and the dynamic change that occurs across related military operations. Military history is replete with accounts of adaptation, hinging on a leader's ability to have uncanny insight into the situation, to be keenly self-aware, and to have a mindset and knowledge that promotes adaptation.

5-31. Adaptability for an individual means having broad and deep knowledge and a good mix of skills and characteristics (see table 5-3). Critical and creative thinking skills are needed when new situations are encountered and the team does not have existing knowledge to use in adaptation.

Table 5-3. Skills and characteristics of adaptability

Skills	*Characteristics*
Quickly assess the situation. Recognize changes in the environment. Identify critical elements of new situation. Apply new skills in unanticipated contexts. Change responses readily. Use multiple perspectives through critical and creative thinking. Avoid oversimplification.	Open-minded. Flexible, Versatile, Innovative. Sees change as an opportunity. Passionate learner. Comfortable in unfamiliar environments. Comfortable with ambiguity. Maintain appropriate complexity in knowledge.

5-32. Adaptability for a team means having a variety of skills within the team to enable adaptation. Adaptability is enhanced when members of the team apply unique knowledge to a problem in new ways. Developing expertise is important to enable adaptable performance later. Having multiple cues to knowledge determines whether atypical, yet useful, knowledge is recalled when needed. Automatic recall can allow greater spare capacity to deal with novel and complex aspects of a problem. Automatic recall, such as pattern recognition, can develop through repeated training beyond performance standards. Being able to adapt depends on the effort ahead of time that goes into developing the capability to adapt.

5-33. While many think of adaptability as a constant good, changing from a known, workable response is not always the best course. Adaptation involves knowing or deciding whether to adapt, what to adapt to, over what timeframe to adapt, and how to adapt. Adaptability is enabled by—

- Recognizing the need for change or recognize a need to take action.
- Knowing the cues that point to real, meaningful differences and cause-effect relationships.
- Having a keen ability to discriminate among environmental cues.
- Having flexible knowledge triggered from different cues. Useful knowledge is likely structured in modular chunks that can recombine in new ways. Understanding the principles and theory behind facts can contribute to novel application of knowledge. This characteristic is cognitive flexibility.
- Seeing multiple sides of an issue and a drive to work toward the best one. Often, multiple sides need integration to derive the best perspective. Openness, seeing opposites, selecting the best of opposing approaches, designing compromise, or resolving contradictions aid integration.
- Thinking in reverse time. This involves being able to think from a desired end state through the prior steps that reach it. It may involve going from constraints or possibilities to figure what is doable, what are plausible goals.
- Handling multiple lines of thought. Involves tracking multiple issues or questions, prioritizing among them, remembering lesser issues while maintaining an overarching perspective, and returning to think about lesser issues when there is time available to think about them.
- Changing perspective. This is referred to as decentering and involves an ability to move away from one's center or viewpoint to overcome thinking obstacles and blind spots.
- Thinking in progressively deeper ways. Involves thinking at the right level of depth and breadth that optimize effort on thinking to match the gravity of the situation.
- Predicting. Involves going beyond first-order or obvious meaning, to broaden thinking to future classes of situations.
- Visualizing and conceptualizing. Involves ability to imagine complex or unusual relationships, possibilities, or unforeseen consequences and relationships.
- Thinking holistically. Involves seeing wholes, sets of relationships and interactions, instead of analytical, decomposed, individual, or isolated parts. Relates to an ability to "see" in dynamics—moving pictures—instead of a static snapshot.
- Mentally simulating what could happen. Mental simulation means to mentally construct and think through a model of a problem, situation, or potential solution to determine important relationships. The process will gauge how much of some action or resource does it take to create a noticeable difference in an outcome?

5-34. To develop adaptability, leaders encourage the following by planning individual or unit events or reinforcing them as they occur during the normal course of collective training or operations:

- Develop sound foundational knowledge and encourage the search for other sources of information. Having a substantial base of knowledge allows leaders to have something ready to apply to new situations and to adjust from the known to the unfamiliar.
- Expand ways of thinking through emphasis on improving critical and creative thinking. Since adaptability opportunities occur in unfamiliar situations, leaders will not have a past answer to apply. Leaders can adapt by thinking through the change using principles of critical and creative thinking. Critical thinking helps make fine-distinctions and connections among concepts, which is useful when analyzing a situation or generating and evaluating solutions.
- Practice with repetition under varied, challenging conditions intentionally selected to prompt adaptability. Practice should allow adequate time for feedback and reflection. Many practice experiences allow leaders to learn about their ability to form situational understanding and the fit of their thought process to multiple problems and the variations that can occur.
- Take advantage of daily events as opportunities for learning, practice, and reflection. Leaders who have a mindset for learning from all activities will be creating knowledge and patterns of thought that can apply to unpredicted situations.
- Create and maintain a supportive culture of innovation, autonomy, and freedom to fail. Learning organizations support the conditions where learning and development will thrive.

Chapter 6
Leader Performance Indicators

6-1. Accurate, descriptive observations of leadership are important to assess performance and provide feedback that produce focused learning. Assessing an individual's performance into the categories of developmental need, meets standard, and strength informs the individual about what needs development or sustainment. It will also provide motivation to develop. The behavior indicators in this chapter provide some general performance measures for varying levels of proficiency for the leader attributes and core leader competencies. Understanding the behavior indicators and observation methods provides a strong base for providing feedback to subordinates.

> **Motivate with High Expectations**
>
> From a command sergeant major:
> We should not just accept normal from our Soldiers, we should instill vitality and flow and high motivation so they can grow and develop and reach their untapped potential. Organizationally, we should not look to be just effective or efficient but shoot for excellence and extraordinary. When it comes to adaptation we should be flourishing and not just coping, and we should look to be flawless in our quality and not just reliable. This approach will not only make Soldiers more excellent in how they do their duties but will create organizations that can operate in any environment, under any conditions and provide extraordinary results.

ACCURATE AND DESCRIPTIVE OBSERVATIONS

6-2. Observing leadership occurs by watching how a leader interacts with others and influences them. Written directives, verbal communications, and leader actions all provide indications of how a leader performs. Raters also learn about their subordinates' leadership by observing reactions to the subordinate among peers, subordinates, and other superiors.

6-3. When observing leadership, these key components ensure observations are accurate and descriptive:

- Plan to take multiple observations over several months or during a rating period. Use both key events and routine operations.
- Make observations based on ADRP 6-22, ADRP 1, and the individual's duty descriptions and performance objectives. Look for patterns of behavior. Seek to confirm initial impressions. Be alert to changes in performance and causes for strengths, inconsistencies, or developmental needs.
- Record important observations immediately for later use in performance and professional growth counseling and for evaluations.
- Consider dimensions on which performance can be differentiated such as the extent of demonstration of a desired behavior, the ability and initiative shown in learning to improve or engage in a desired behavior, and the extent and duration of effects that the behavior has on individual or unit performance.

APPLICATION OF THE PERFORMANCE INDICATORS

6-4. Performance indicators are grouped according to the doctrinal leadership requirements model in categories of leader attributes (character, presence, and intellect) and leader competencies (lead, develop, and achieve). The performance indicators provide three levels of proficiency: a developmental need, the standard, and a strength. For developmental purposes, these three categories are sufficient and apply across cohorts. A developmental need is identified as a specific need for development when the observed individual does not demonstrate the leader competency. Strength indicators are associated with successful performance of a leader attribute or competency. Strengths include a consistent pattern of natural talents, knowledge gained through learning, and skills acquired through practice and experience.

6-5. While comparing observations against the leader performance indicators, determine the level of proficiency of the observed leader: first review the behavior that appears in the center column of tables 6-1 through 6-6 on pages 6-3 through 6-8—this represents the standard for leader performance. A leader demonstrating quality leadership to standard will exhibit decisions and actions described in the center column. The column on the left describes performance indicating a developmental need (individual falls short of the standard), while the column to the right describes performance indicating a strength (individual exceeds the standard).

6-6. Understanding the competencies and attributes in the Army leadership requirements model is essential to make careful and accurate observations of a subordinate's performance and evaluation of potential. The core leader competencies include how Army leaders lead people; develop themselves, their subordinates, and organizations; and achieve the mission. The competencies are the most outwardly visible signs of a leader's performance. Leader attributes are inward characteristics of the individual that shape the motivations for actions and bearing, and how thinking affects decisions and interactions with others.

Competency Development

From an Army Civilian supervisor:

At midpoint and annual performance reviews, I hold stakeholder dialogues with individual employees. I ask them to give me examples of where they demonstrated leadership, and I ask them what I can do better, let them know what they do well, should keep doing, or start doing. Each subordinate selects one competency from the Army leadership requirements model to improve throughout their rating period.

6-7. The information in these tables is illustrative of a focus on core leadership characteristics. The Army adoption of a core attribute and competency model means that no list will be comprehensive of all performance requirements for any leader. Each rater, counselor, mentor, or trainer will need to expand the set to specific duty or functional requirements. They should be able to apply the ideas to specific performance objectives designated for individuals that exceed the core leadership requirements.

ATTRIBUTE CATEGORIES

6-8. The leader attributes are presented in three categories: character, presence, and intellect.

Character

6-9. ADRP 6-22 defines character as factors internal and central to a leader, which make up an individual's core and are the mindset and moral foundation behind actions and decisions. Leaders of character adhere to the Army Values, display empathy and the Warrior Ethos/Service Ethos, and practice good discipline. See table 6-1.

Table 6-1. Framing the Army Values, empathy, Warrior/Service ethos, and discipline

DEVELOPMENTAL NEED	*STANDARD*	*STRENGTH*
	ARMY VALUES	
Inconsistently demonstrates: loyalty, duty, respect, selfless service, honor, integrity, and personal courage. Demonstrates these with more than occasional lapses in judgment.	Consistently demonstrates: loyalty, duty, respect, selfless service, honor, integrity, and personal courage.	Models loyalty, duty, respect, selfless service, honor, integrity, and personal courage. Promotes the associated principles, standards, and qualities in others.
	EMPATHY	
Exhibits resistance or limited perspective on the needs of others. Words and actions communicate lack of understanding or indifference. Unapproachable and disinterested in personally caring for Soldiers.	Demonstrates an understanding of another person's point of view. Identifies with others' feelings and emotions. Displays a desire to care for Soldiers, Army Civilians, and others.	Attentive to other's views and concerns. Takes personal action to improve the situation of Soldiers, Army, Civilians, family members, local community, and even that of potential adversaries. Breaks into training, coaching, or counseling mode when needed and role models empathy for others.
	WARRIOR ETHOS/SERVICE ETHOS	
Inconsistently demonstrates the spirit of the profession of arms. Downplays the importance of this sentiment.	Demonstrates the spirit of the profession of arms and commitment to the mission, to never accept defeat, to persevere through difficulties, and to always support fellow Soldiers.	Models the spirit of the profession of arms. Instills this behavior in subordinates and others.
	DISCIPLINE	
Fails consistently to adhere to rules, regulations, or standard operating procedures.	Demonstrates control of one's own behavior according to Army Values and adheres to the orderly practice of completing duties of an administrative, organizational, training, or operational nature.	Demonstrates discipline in one's own performance and encourages others to follow good practices of discipline as well. As situations call for it, enforces discipline when others fail to adhere to Army Values or to other standard practices.

Presence

6-10. Presence is how others perceive a leader based on the leader's appearance, demeanor, actions, and words. Leaders with presence demonstrate military and professional bearing, fitness, confidence, and resilience. See table 6-2 on page 6-4.

Table 6-2. Framing presence

DEVELOPMENTAL NEED	*STANDARD*	*STRENGTH*
	MILITARY AND PROFESSIONAL BEARING	
Inconsistently projects a professional image of authority. Actions lack a commanding presence. Allows professional standards to lapse in personal appearance, demeanor, actions, and words.	Possesses a commanding presence. Projects a professional image of authority. Demonstrates adherence to standards.	Models a professional image of authority. Commanding presence energizes others. Exemplifies adherence to standards through appearance, demeanor, actions, and words.
	FITNESS	
Physical health, strength, or endurance is not sufficient to complete most missions. Fitness level unable to support emotional health and conceptual abilities under prolonged stress.	Displays sound health, strength, and endurance that support emotional health and conceptual abilities under prolonged stress.	Models physical health and fitness. Strength and endurance supports emotional health and conceptual abilities under prolonged stress. Energetic attitude conveys importance of fitness to others.
	CONFIDENCE	
Inconsistently displays composure or a calm presence. Allows a setback to derail motivation. Displays a less than professional image of self or unit.	Displays composure, confidence, and mission-focus under stress. Effectively manages own emotions and remains in control of own emotions when situations become adverse.	Projects self-confidence and inspires confidence in others. Models composure, an outward calm, and control over emotions in adverse situations. Manages personal stress, and remains supportive of stress in others.
	RESILIENCE	
Slowly recovers from adversity or stress. Inconsistently maintains a mission or organizational focus after a setback.	Recovers from setbacks, shock, injuries, adversity, and stress while maintaining a mission and organizational focus.	Quickly recovers from setbacks. Focuses on the mission and objectives during shock, injuries, and stress. Maintains organizational focus despite adversity. Learns from adverse situations and grows in resilience.

Intellect

6-11. Intellect is comprised of the mental tendencies or resources that shape a leader's conceptual abilities and affect a leader's duties and responsibilities. Leaders with high intellect are mentally agile, good at judgment, innovative, tactful around others, and expert in technical, tactical, cultural, geopolitical, and other relevant knowledge areas. See table 6-3.

Table 6-3. Framing intellect

DEVELOPMENTAL NEED	*STANDARD*	*STRENGTH*
	MENTAL AGILITY	
Inconsistently adapts to changing situations. Attends to immediate conditions and surface outcomes when making decisions. Hesitates to adjust an approach.	Demonstrates open-mindedness. Recognizes changing conditions and considers second- and third-order effects when making decisions.	Models a flexible mindset and anticipates changing conditions. Engages in multiple approaches when assessing, conceptualizing, and evaluating a course of action.
	SOUND JUDGMENT	
Inconsistently demonstrates sound assessment of situations. Hesitates in decisionmaking when facts not available. Forms opinions outside of sensible information available.	Demonstrates sound decisionmaking ability. Shows consideration for available information, even when incomplete.	Models sound judgment. Engages in thoughtful assessment. Confidently makes decisions in the absence of all of the facts.
	INNOVATION	
Relies on traditional methods when faced with challenging circumstances.	Offers new ideas when given the opportunity. Provides novel recommendations when appropriate.	Consistently introduces new ideas when opportunities exist to exploit success or mitigate failure. Creatively approaches challenging circumstances and produces worthwhile recommendations.
	INTERPERSONAL TACT	
Demonstrates lapses in self-awareness when interacting with others. Misses cues regarding others perceptions, character and motives. Presents self inappropriately or not tactfully	Maintains self-awareness of others perceptions and changes behaviors during interactions accordingly.	Demonstrates proficient interaction with others. Effectively adjusts behaviors when interacting with others. Understands character and motives of others, and modifies personal behavior accordingly.
	EXPERTISE	
Demonstrates uncertainty or novice proficiency in technical aspects of position. Inconsistently applies competence of joint, cultural, and geopolitical knowledge. Displays indifference toward expanding knowledge or skill set	Possesses facts and understanding of joint, cultural, and geopolitical events and situations, Seeks out information on systems, equipment, capabilities, and situations. Expands personal knowledge of technical, technological, and tactical areas.	Demonstrates expert-level proficiency with technical aspects of their position. Demonstrates understanding of joint, cultural, and geopolitical knowledge. Shares knowledge of technical, technological, and tactical systems to subordinates and others.

CORE LEADER COMPETENCY CATEGORIES

6-12. The core leader competencies are presented in three categories: lead, develop, and achieve.

Lead

6-13. Leaders set goals and establish a vision, motivate or influence others to pursue the goals, build trust to improve relationships, communicate and come to a shared understanding, serve as a role model by displaying character, confidence, and competence, and influence outside the chain of command. See table 6-4 on page 6-6.

Table 6-4. Framing leads

DEVELOPMENTAL NEED	*STANDARD*	*STRENGTH*
	LEADS OTHERS	
Inconsistently demonstrates influence techniques. Fails to monitor risk factors affecting others. Allows mission priority to affect subordinate morale, physical condition, or safety adversely. Hesitates to act when risk factors escalate.	Influences others effectively. Assesses and routinely monitors effects of task execution on subordinate welfare. Monitors conditions of subordinate morale and safety. Implements appropriate interventions when conditions jeopardize mission success. Assesses and manages risk.	Demonstrates full range of influence techniques. Continually assesses and monitors mission accomplishment and Soldier welfare. Attends to subordinate morale, physical condition, and safety. Implements interventions to improve situations. Assesses and mitigates prudent risk to maximize potential for success.
	BUILDS TRUST	
Inconsistently demonstrates trust. Displays respect differently to some without justification. Takes no actions to build rapport or trust with others. Fails to address problems caused by team members who undermine trust. Fails to follow through on intentions, undermining the trust others would have in this leader.	Establishes trust by demonstrating respect to others and treating others in a fair manner. Uses common experiences to relate to others and build positive rapport. Engages others in activities and sharing of information that contribute to trust.	Demonstrates trust in others when encountering new or unfamiliar situations. Bases trust on a thorough understanding of trustworthiness of others and self. Understands how much trust to project and grant to others. No hesitation in addressing problems that undermine trust.
	EXTENDS INFLUENCE BEYOND THE CHAIN OF COMMAND	
Inconsistently demonstrates understanding of indirect influence. Misses or passively acts on opportunities to build trusting relationships outside the organization.	Demonstrates understanding of conditions of indirect influence. Builds trust to extend influence outside the organization. Displays understanding of the importance of building alliances.	Demonstrates effective use of indirect influence techniques. Establishes trust to extend influence outside the chain of command. Proactively builds positive relationships inside and outside the organization to support mission accomplishment.
	LEADS BY EXAMPLE	
Demonstrates conduct inconsistent with the Army Values. Displays a lack of commitment and action. Remains unaware of or unconcerned about the example being set.	Demonstrates an understanding of leader attributes and competencies. Recognizes the influence of personal behavior and the example being set. Displays confidence and commitment when leading others.	Models sound leader attributes and competencies. Exemplifies the Warrior Ethos through actions regardless of situation. Demonstrates competence, confidence, commitment, and an expectation of such behavior in others.
	COMMUNICATES	
Misunderstands or fails to perceive nonverbal cues. Ideas not well organized or easily understandable. Speaks without considering listener interest. Information dissemination is inconsistent or untimely.	Chooses appropriate information-sharing strategy before communicating. Conveys thoughts and ideas appropriately. Disseminates information promptly. Provides guidance and asks for a brief back or confirmation.	Uses verbal and nonverbal means to maintain listener interest. Adjusts information-sharing strategy based on operating conditions. Ensures prompt information dissemination to all levels. Avoids miscommunication through verifying a shared understanding.

Develop

6-14. Leaders foster teamwork; express care for individuals; promote learning; maintain expertise, skills, and self-awareness; coach, counsel and mentor others; foster position development, and steward the profession of arms. See table 6-5.

Table 6-5. Framing develops

DEVELOPMENTAL NEED	*STANDARD*	*STRENGTH*
	CREATES A POSITIVE ENVIRONMENT/FOSTERS ESPRIT DE CORPS	
Demonstrates negative expectations and attitudes not conducive to a productive work environment. Focuses primarily on task accomplishment. Fosters an expectation of zero-defects. Holds honest mistakes against subordinates.	Promotes expectations and attitudes conducive to a positive work environment. Demonstrates optimism and encourages others to develop and achieve. Provides coaching, counseling and mentoring to others.	Exemplifies a positive attitude and expectations for a productive work environment. Conveys a priority for development in the organization. Encourages innovative, critical, and creative thought. Uses lessons learned to improve organization.
	PREPARES SELF	
Reluctant to accept responsibility for learning. Downplays feedback. Acts on information without regard to source, quality, or relevance. Ineffectively transfers new information into knowledge.	Accepts responsibility for learning and development. Evaluates and incorporates feedback. Analyzes and organizes information to create knowledge. Focuses on credible sources of information to improve personal understanding.	Seeks feedback. Seeks learning opportunities to improve self. Demonstrates knowledge management proficiency. Integrates information from multiple sources; analyzes, prioritizes, and utilizes new information to improve processes.
	DEVELOPS LEADERS	
Disinterested in motivating and enabling the growth of others. Focuses on the task without consideration of improving organizational effectiveness.	Demonstrates willingness to motivate and help others grow. Provides coaching, counseling and mentoring. Builds team skills and processes to improve individuals and the organization.	Seizes opportunities to teach, coach and mentor. Fosters position development and enrichment. Knows subordinates and prepares them for new positions. Improves unit productivity.
	STEWARDS THE PROFESSION	
Fails to extend assistance to others or other units. Disregards oversight of the tracking and use of resources. Fails to improve subordinates for subsequent assignments and fails to take steps to leave the organization in equal or better condition than when this leader arrived.	Supports developmental opportunities of subordinates. Takes steps to improve the organization. Carefully manages resources of time, equipment, people, and money.	Applies a mindset that looks to strengthen the profession of arms into the future. Assumes some risk to forego some short-term or personal gains in favor of improving one's own organization, other units, and other individuals. Cooperates by providing more assistance to others than expected to receive in return.

Achieve

6-15. Leaders achieve by setting priorities, organizing taskings, managing resources, developing thorough and synchronized plans, executing plans to accomplish the mission, and achieving goals See table 6-6 on page 6-8.

Table 6-6. Framing achieves

DEVELOPMENTAL NEED	*STANDARD*	*STRENGTH*
	GETS RESULTS	
Demonstrates a limited understanding of supervising, managing, monitoring, and controlling priorities of work. Hasty prioritization and planning lead to incomplete guidance and direction.	Prioritizes, organizes, and coordinates taskings for others. Plans for expected setbacks and enacts appropriate contingencies when needed. Monitors, coordinates and regulates subordinate actions but allows subordinates to accomplish the work.	Utilizes other competencies to accomplish objectives. Demonstrates understanding of supervising, managing, monitoring, and controlling of priorities of work. Reflects on end state before issuing guidance. Provides subordinates autonomy to accomplish the work.

Chapter 7

Learning and Development Activities

7-1. This chapter is a guide for all Army leaders to develop themselves or to develop others. These activities follow the same organization as the leader competencies found in the Army leadership requirement model (see ADRP 6-22): ten leader competencies grouped in the categories of lead, develop, and achieve. Each developmental action listed in this chapter follows the same format: strength and need indicators, underlying causes, and activities for feedback, study, and practice. The indicators provide ways to understand leader actions and confirm aspects of each leader behavior as a strength or a developmental need. The underlying causes provide more information on what the root cause may be for a developmental need. The tables provide three options for developmental action: feedback, study, and practice.

7-2. To best use this chapter's information, first identify the competencies and behaviors for developmental focus. An individual may already have an IDP that documents goals and a plan for development or have an idea of what leadership competency or skill to develop. A coach, rater, counselor, or mentor can use this material to help focus leaders or subordinates on specific developmental goals. Different sources and events inform the identification of developmental goals for competencies and behaviors as illustrated in table 7-1.

Table 7-1. Identification of developmental goal

Source or event providing identification of developmental goal	***Example***
Interest to the developing leader	'I want to get better at setting a positive climate that encourages subordinates to promote development.'
360° assessment and feedback report and/or coaching session	Communication skills make up the lowest assessed area relative to all other areas.
Performance evaluation and developmental growth counseling session	'You are good at motivating your Soldiers, you could grow into an even better leader by learning to better integrate tasks, resources and priorities to achieve results.'
Mentor's advice	'To move to the next level you could learn additional ways to operate with others outside the Army and to extend influence.'
Self-realization during institutional education course	'My fellow students generally seem more knowledgeable than me about world affairs affecting our Army.'
Counterpart feedback received during a training center rotation or home station training	'Under stress you are overly directive which doesn't align fully with mission command; you could learn to use commitment-building actions to expand your toolkit of influence.'

Tip: When considering learning and developmental activities, some may automatically think of taking a formal training course or reading. While these may be helpful, leaders are encouraged to select developmental activities that fit with personal learning-style preferences and situation. It is important to think through personal and career goals when deciding on a developmental activity.

CAPABILITY EVALUATION AND EXPANSION

7-3. To start using any of the developmental action tables, there are a few guidelines to consider for the most benefit from them. Each section is designed to help a leader understand and act on strengths as well as developmental needs. Some leaders may experience greater growth by focusing on improving strengths rather than focusing on developmental needs or using strengths to address developmental needs.

CAPABILITY EVALUATION—STRENGTH AND NEED INDICATORS AND UNDERLYING CAUSES

7-4. Evaluating capabilities involves identifying personal practices that support or hinder successful performance. Each table includes diagnostics to enable evaluation of how well an individual is doing on that behavior and provides examples of why they may or may not be excelling. Consider if the strength and need indicators represent personal behaviors. Each diagnostic section includes:

- Strength Indicators: Behaviors and actions that contribute to or support successful performance.
- Need Indicators: Behaviors and actions that reduce or hinder successful performance.
- Underlying Causes: Examples of why an individual may not be excelling at a particular leader behavior.

CAPABILITY EXPANSION—FEEDBACK, STUDY, AND PRACTICE

7-5. To build on an individual's current level, review the developmental activities for each capability area and personalize them. Table 7-2 outlines methods to engage in developmental activities. The developmental activities include:

- Feedback. Sources and methods for obtaining feedback to guide self-development efforts.
- Study. Topics and activities to learn more about a behavior.
- Practice. Actions to improve skill and comfort in performing a leader behavior.

Table 7-2. Methods to implement developmental activities

Developmental Step	*Options to take*	*Method*
Feedback	Ask for feedback…	From others about how you are doing with specific issues and areas of performance.
	Gain support…	From peers, colleagues, friends, or other people who can provide encouragement or recognize success.
	Consult…	With friends, supervisors, peers, subordinates, coaches, mentors, or other professionals to give advice on strengths or areas of concern.
Study	Observe…	Other leaders, professionals, and similar organizations. Note the most or least effective behaviors, attributes, and attitudes.
	Make time to reflect on…	Personal or situational characteristics that relate to the strength or need. Consider alternative perspectives.
	Read…	Books, articles, manuals, and professional publications.
	Investigate…	A topic through internet or library searches, gathering or asking questions, and soliciting information and materials.
Practice	Practice…	A skill or behavior that needs improvement in a work situation or away from the unit.
	Participate in training…	Including Army schools, unit training programs, outside seminars, degree programs, and professional certifications.
	Teach…	A skill you are learning to someone else.
	Accept an opportunity…	That stretches personal abilities, such as giving presentations, teaching classes, volunteering for special duty assignments, position cross-training, and representing the unit at meetings.
	Explore off-duty events…	Such as leading community groups, trying a new skill in a volunteer organization, or presenting to schools and civic organizations.

DEVELOPMENTAL ACTIVITIES

7-6. Table 7-3 will assist an individual in determining where to start development activities. If the individual needs greater understanding to direct development, they should first seek feedback and follow with study and practice. If a developmental need is understood but knowing how to address that need is unknown, the individual should start with study and follow with practice. If the only unknown is what to practice, then the individual should focus on the practice activities. Applying the if-then logic in table 7-3 to each developmental goal will help individuals get the most from their development efforts.

Table 7-3. Evaluation model

If...	*Then...*
I need more insight into how well I am demonstrating a competency or component and what I can do to improve...	I should seek *Feedback*. Feedback is an opportunity to gain information about how well you are doing. Feedback can include direct feedback, personal observations, analysis of response patterns, and acknowledgement of outcomes.
I need to gain or expand my understanding of theory, principles, or knowledge of a leader competency or component...	I should *Study*. Study facilitates an intellectual understanding of the topic. Study can include attending training courses, reading, watching movies, observing others on duty, and analyzing various sources of information.
I need more experience to build or enhance my capability through opportunities to perform a leader competency or component...	I should *Practice*. Practice provides activities to convert personal learning into action. Practice includes engaging in physical exercises, team activities, rehearsals, and drills.

7-7. View all suggestions for developmental activities through a personal lens. The following questions are sample questions to ask when refining a development activity to fit personal needs and situation. Depending on the chosen activity, other considerations may be important too. Be willing to take risks and choose activities outside personal comfort zones to challenge yourself and accelerate development.

DETERMINING DEVELOPMENTAL ACTIVITIES

Answer these to focus selection of appropriate developmental activities.

Developmental Activity: What do I want to do?

Desired Outcome: What do I hope to achieve?

Method: How am I going to do this? What resources do I need?

Time available: When will I do this? How will I monitor progress (such as identifying and monitoring milestones, rewarding success, or identifying accountability partners)?

Limits: What factors will affect or hinder successful implementation of this activity?

Controls: What can minimize or control the factors that hinder implementation of this activity?

7-8. Use table 7-4 starting on page 7-4 to locate appropriate developmental activities. The MSAF 360° feedback reports detail individual or unit strengths and developmental needs. Learners and coaches will select a few actions at a time to guide development.

Table 7-4. Leadership competencies and actions listing

Competency	*To find developmental activities for...*	*Go to...*
Leads others	Uses appropriate methods of influence to energize others.	Table 7-5.
	Provides purpose.	Table 7-6.
	Enforces standards.	Table 7-7.
	Balances mission and welfare of followers.	Table 7-8.
Builds trust	Sets personal example for trust.	Table 7-9.
	Takes direct actions to build trust.	Table 7-10.
	Sustains a climate of trust.	Table 7-11.
Extends Influence	Understands sphere, means, and limits of influence.	Table 7-12.
	Negotiates, builds consensus, and resolves conflict.	Table 7-13.
Leads by example	Displays Army Values.	Table 7-14.
	Displays empathy.	Table 7-15.
	Exemplifies the Warrior Ethos/Service Ethos.	Table 7-16.
	Applies discipline.	Table 7-17.
	Leads with confidence in adverse situations.	Table 7-18.
	Demonstrates tactical and technical competence.	Table 7-19.
	Understands and models conceptual skills.	Table 7-20.
	Seeks diverse ideas and points of view.	Table 7-21.
Communicates	Listens actively.	Table 7-22.
	Creates shared understanding.	Table 7-23.
	Employs engaging communication techniques.	Table 7-24.
	Sensitive to cultural factors in communication.	Table 7-25.
Creates a positiv eenvironment/ esprit de corps	Fosters teamwork, cohesion, cooperation, and loyalty (esprit de corps).	Table 7-26.
	Encourages fairness and inclusiveness.	Table 7-27.
	Encourages open and candid communications.	Table 7-28.
	Creates a learning environment.	Table 7-29.
	Encourages subordinates.	Table 7-30.
	Demonstrates care for follower well-being.	Table 7-31.
	Anticipates people's duty needs.	Table 7-32.
	Sets and maintains high expectations for individuals and teams.	Table 7-33.
Prepares self	Maintains mental and physical health and well-being.	Table 7-34.
	Expands knowledge of technical, technological, and tactical areas.	Table 7-35.
	Expands conceptual and interpersonal capabilities.	Table 7-36.
	Analyzes and organizes information to create knowledge.	Table 7-37.
	Maintains relevant cultural awareness.	Table 7-38.
	Maintains relevant geopolitical awareness.	Table 7-39.
	Maintains self-awareness.	Table 7-40.
Develops others	Assesses developmental needs of others.	Table 7-41.
	Counsels, coaches, and mentors.	Table 7-42.
	Facilitates ongoing development.	Table 7-43.
	Builds team skills and processes.	Table 7-44.
Stewards the profession	Supports professional and personal growth.	Table 7-45.
	Improves the organization.	Table 7-46.

Table 7-4. Leadership competencies and actions listing (continued)

Gets results	Prioritizes, organizes, and coordinates taskings.	Table 7-47.
	Identifies and accounts for capabilities and commitment to task.	Table 7-48.
	Designates, clarifies, and deconflicts duties and responsibilities.	Table 7-49.
	Identifies, contends for, allocates, and manages resources.	Table 7-50.
	Removes work obstacles.	Table 7-51.
	Recognizes and rewards good performance.	Table 7-52.
	Seeks, recognizes, and takes advantage of opportunities.	Table 7-53.
	Makes feedback part of work processes.	Table 7-54.
	Executes plans to accomplish the mission.	Table 7-55.
	Identifies and adjusts to external influences.	Table 7-56.

LEADS OTHERS

7-9. Leaders motivate, inspire, and influence others to take initiative, work toward a common purpose, accomplish critical tasks, and achieve organizational objectives. Influence focuses on motivating and inspiring others to go beyond their individual interests and focus on contributing to the mission and the common good of the team. The leads others competency has four components:

- Uses appropriate methods of influence to energize others.
- Provides purpose.
- Enforces standards.
- Balances mission and welfare of followers.

USES APPROPRIATE METHODS OF INFLUENCE TO ENERGIZE OTHERS

7-10. Army leaders can draw on a variety of techniques to influence others ranging from obtaining compliance to building commitment to a cause or organization. Specific techniques for influence fall along a continuum including pressure, legitimate requests, exchange, personal appeals, collaboration, rational persuasion, apprising, inspiration, participation, and relationship building (see ADRP 6-22 for more information). To succeed in creating true commitment, leaders determine the proper influence technique based on the situation and individuals involved. Keep in mind that the effects of influence are not often instantaneous. It may take time before seeing positive, enduring results (see table 7-5).

Table 7-5. Uses appropriate methods of influence to energize others

Strength Indicators	*Needs Indicators*
Assesses the situation and determines the best influence technique to foster commitment. Considers the mission when exerting influence. Uses positive influence to do what is right. Uses pressure only when the stakes are high, time is short, and attempts at achieving commitment are not successful.	Uses a single or limited number of influence techniques for all influence without consideration of the circumstances or individuals involved. Coerces or manipulates the situation to achieve personal gain. Subordinates return several times to clarify what to do.
Underlying Causes	
Lack of understanding of the individuals to be influenced (values, needs, or opinions). Lack of awareness of likely effects (advantages and disadvantages) of influence techniques on others. Does not match the appropriate influence technique to the individual and does not factor in contextual causes (such as high operational tempo, significant stress, speed of situational changes). Focus on personal gain and accomplishment rather than doing what is right for the Army and the unit. Too forceful or not forceful enough when applying influence techniques. Belief that collaborative or rational approaches to gaining desired behavior weakens personal authority.	
Feedback	Conduct reviews with team members and subordinates; listen for clues on the style and method of influence that works best for the team. Periodically speak with subordinates to ensure that your influence creates a positive environment and is in line with Army expectations. Complete a self-assessment tool to understand the way you operate and its effect on your approach and style of influence. Talk to subordinates about what influence they find most effective with certain tasks.
Study	Proactively seek information to understand what is important to those you are trying to influence. Create the message to address the stakeholder's key needs and concerns. Identify the appropriate influence technique by analyzing the criticality and time available for obtaining the desired behavior and the disposition of those you seek to influence. Observe and analyze different ways that you influence others noting what seems to be most effective for different tasks, situations, and individual dispositions. Consider the everyday stresses, obligations, interests, values, and dispositions of those whom you are trying to influence. Choose influence techniques to produce the best results under these circumstances. Ensure the influence technique aligns with the Army Values, ethical principles, and the Uniform Code of Military Justice. When leading the team to mission accomplishment, use the least coercive and most cooperative influence techniques under the circumstances to build and sustain task ownership and enhanced motivation. Stay persistent, influence is not instantaneous and may require repeated action.
Practice	Review influence techniques (see ADRP 6-22). Identify methods that are strengths and those that cause struggle. Create an action plan to develop the full set. Identify when to use compliance-focused influence (based primarily on authority) and when to use commitment-focused influence (seeks to change attitudes and beliefs). Contact former superiors about ways they handled conflict and influence. Ask what worked best and common mistakes that occur in a high stress situation. Research available methods of group collaboration. Teams can have widely different dynamics so a better understanding of different methods will help adaptation. Explore personal beliefs and assumptions about being a leader, authority, and senior-subordinate relationships. Consider how beliefs affect the methods of influence used. Access the Virtual Improvement Center to complete: Making Influence Count, Motivating through Rewards, Enabling Subordinates Using Mission-Focused Delegation, Beyond People Skills: Leveraging Your Understanding of Others.

Provides Purpose

7-11. Establishing and imparting a clear sense of intent and purpose serves as a catalyst to getting work done by providing a distinct path forward. Oftentimes, with a firm sense of purpose, the result is easier to reach. Defining a clear sense of purpose can be difficult, as it requires thinking about the objective or task from a macro-level before getting involved in implementing the details. However, developing clear intent and purpose can provide substantial benefits by clarifying required actions and resources as well as aligning the efforts of the team (see table 7-6).

Table 7-6. Provides purpose

Strength Indicators	***Need Indicators***
Determines goals or objectives. Translates task goals and objectives into a sequenced action plan. Restates the mission so that it resonates with the unit and is understood easily. Communicates clear instructions that detail each process step through task completion; provides guidance as needed throughout the process. Focuses on the most important aspects of a mission to emphasize priorities and align efforts. Empowers authority to the lowest level possible.	Restates the mission in a manner that subordinates do not understand. Fails to provide strong, clear direction to team members and subordinates. Keeps authority and decisionmaking centralized. Keeps subordinates in the dark; fails to recognize the need to understand the goal. Does not set a standard for expected contributions to the team. Subordinates often must come back several times to clarify task goals.
Underlying Causes	
Has not formed a clear purpose and intent in own mind. Does not fully understand the objectives of a given mission or task. Superiors failed to articulate the mission clearly. Difficulty in expressing intent and purpose in terms others can easily understand and visualize. Uncomfortable with relinquishing personal control and authority over the task or unit. Not confident in subordinates' abilities to make decisions and achieve the purpose and intent. Fails to adapt to complexity, ambiguity, or stress of a situation.	
Feedback	Ask subordinates if the purpose and intent are clear. Have them backbrief the purpose and intent. Ask what could facilitate their understanding of what you are trying to convey. Talk to team members about the clarity of their task assignments. Do they understand how the work they complete contributes to the organizational goals? Listen to feedback from superiors, peers, and subordinates about your communication skills. Determine which are effective or ineffective in imparting the mission purpose and intent.
Study	Study subordinates' reactions when first establishing mission goals and purpose. Do their facial expressions and body language convey understanding or confusion? Identify a unit member who is a strong planner and mission briefer. Watch their actions. How do these actions compare to what you typically do? Study how other leaders impart clear purpose and intent to subordinates. Discuss the thought process for identifying, planning, and communicating purpose and intent. Examine organizational or commanders' vision statements or past operations orders. Note how intent, purpose, and communicated vision are expressed. How might they have been expressed more effectively? Access the Virtual Improvement Center to complete: Clarifying Roles; Creating and Supporting Challenging Assignments; Motivating through Rewards; Creating and Promulgating a Vision of the Future; Rapid Team Stand-up: How to Build Your Team ASAP; Enabling Subordinates Using Mission-Focused Delegation.
	ASAP: as soon as possible

Table 7-6. Provides purpose (continued)

Practice	When receiving a mission, brief-back the mission and higher commander's intent in your own words to ensure personal understanding of what to accomplish. When planning a task or mission, begin by visualizing and drafting a written description of the end-state that you want to achieve. When giving a mission to the team, create a detailed plan of execution outlining responsibilities. Show how individual responsibilities relate to the purpose and desired outcomes of the overall mission. Create an open environment where subordinates feel comfortable approaching you to discuss and brainstorm how to complete tasks and missions.

Enforces Standards

7-12. To lead others and gauge correct performance of duties, the Army has established standards for military activities. Standards are formal, detailed instructions to describe, measure, and achieve. To use standards effectively, leaders should explain the standards that apply to the organization and give subordinates the authority to enforce them (see table 7-7).

Table 7-7. Enforces standards

Strength Indicators	***Need Indicators***
Reinforces the importance and role of standards. Explains the standards and their significance. Prioritizes unit activities to ensure not everything is a number one priority. Ensures tasks meet established standards. Recognizes and takes responsibility for poor performance and addresses it properly. Sets attainable milestones to meet the standard.	Focuses on too many priorities at one time. Ignores established individual and organizational standards. Overlooks critical errors instead of dealing with them. Blames substandard outcomes on others.
Underlying Causes	
Does not know or accept established standards. Does not want to be viewed by subordinates as too demanding. Poor self-discipline in meeting standards and setting a personal example. Is unable to handle the complexity of tracking and enforcing standards for multiple tasks or individuals. Does not follow-up on task delegations to ensure standards are met.	
Feedback	Obtain objective and subjective assessments of individual and collective performance. Compare to established standards to identify performance strengths and developmental needs. Engage organizational leaders in discussion and examination of performance standards, including how well standards are communicated, known, enforced, and achieved.
Study	Learn established Army standards for individual and collective tasks expected of the unit. Research how successful leaders have established, communicated, monitored, and enforced individual and collective standards. Consult with superiors about organizational standards most critical to attaining the higher commander's vision and intent. Consider how these standards pertain to the unit.
Practice	When assigning tasks, explicitly state the standard of performance and expectations. When assigning performance standards, explain why the standard is essential for organizational success. Set the tone when involved with any individual or group task. Make sure you are always giving your best effort and providing an example for the team to follow Recognize team members who exemplify the standards you are trying to reinforce. Identify individuals who repeatedly fail to achieve performance standards; address appropriately. When receiving a mission, verify the standard expected—is it appropriate or necessary?.

Balances Mission and Welfare of Followers

7-13. Team welfare is vital to completing a mission while maintaining morale. Taking care of followers will allow creation of a closer working relationship. Leaders must be able to keep an eye on the mission while being cognizant of and caring for the people working for them (see table 7-8).

Table 7-8. Balances mission and welfare of followers

Strength Indicators	***Need Indicators***
Regularly assesses mission effects on the mental, physical, and emotional well-being of subordinates. Checks-in with team members and subordinates to monitor morale and safety. Provides appropriate relief when difficult conditions risk jeopardizing subordinate success. Builds a cohesive team moving in one direction to achieve common goals. Offers support and resources when a team member seems unnecessarily overloaded.	Ignores the risks of overexerting subordinates. Visibly shows discouragement or disgust when morale struggles due to workload. Is insensitive to signs of high stress or diminishing morale. Does not weigh the importance of the mission against adverse effects on stress, morale, and welfare.
Underlying Causes	
Has tunnel vision regarding completion of the mission; believes in mission accomplishment at almost any cost or does not consider the cost. Is overtaxed or fatigued and becomes too focused on own needs rather than those of the organization. Refusal to delegate tasks for fear of failure; does not see the developmental opportunities. Excessively concerned with personal achievement; avoids negative performance feedback. Generally unsympathetic towards the needs of subordinates.	
Feedback	Gather feedback on mission demands and member welfare using face-to-face interaction. This will give a complete reflection of their status. Seek counsel from a mentor or trusted advisor when dealing with a difficult situation. Have them guide you and provide insight into possible next steps. Discuss proposed missions with other unit leaders to assess the adverse effects of mission execution on the welfare of unit members. Have mental health professionals survey the organization for evidence of excessive stress. Obtain summary information and recommendations for reducing stress levels.
Study	Investigate activities and methods of relief used to counter stress. See what has worked well for other leaders and what could be done better. Regularly assess and document both team and individual morale. Identify the greatest sources of stress for the organization. Look for methods of reducing the stress. Research signs of stress so that you can recognize a problem before it becomes an issue. Learn the symptoms and effects of post-traumatic stress disorder so you can identify it and obtain appropriate help for unit members suffering from it. Observe or consider a leader who succeeded in balancing severe demands or stresses faced by the unit with member welfare. How did that leader do it? What types of behaviors and methods can you model to ensure your success? Consider what messages your own behavior sends about balancing personal welfare and mission requirements. Access the Virtual Improvement Center to complete: Out of Time: Managing Competing Demands.
Practice	Observe daily subordinate morale. Are they struggling with the workload? Is it affecting group morale? Brainstorm with other unit leaders possible solutions to team members' workloads. Take advantage of opportunities to give subordinates time off when the mission permits. Weigh proposed missions to compare the importance of the intended outcomes against the costs they are likely to impose on the members who will perform them. Look for ways to minimize costs while still obtaining benefits.

BUILDS TRUST

7-14. Trust is essential to all effective relationships, particularly within the Army. Trust facilitates a bond between Soldiers, leaders, the Army, and the Nation that enables mission success.

7-15. Building trust forms on the bedrock of mutual respect, shared understanding, and common experiences. For teams and organizations to function at the highest level, a climate of trust needs to exist. Leaders create a climate of trust by displaying consistency in their actions, and through relationship-building behaviors such as coaching, counseling, and mentoring. This competency has three components:

- Sets personal example for trust.
- Takes direct actions to build trust.
- Sustains a climate of trust.

SETS PERSONAL EXAMPLE FOR TRUST

7-16. Leaders exhibit their beliefs about trust in their actions and behaviors. Setting a personal example inspires those around them to act in the same manner. The actions a leader models to subordinates communicates the values of the leader and the unit. Setting a personal example for trust should be consistent, and is the most powerful tool a leader has to shape the climate of the organization (see table 7-9).

Table 7-9. Sets personal example for trust

Strength Indicators	***Need Indicators***
Follows through on commitments and promises. Presents the truth, even if unpopular or difficult. Protects and safeguards confidential information. Admits mistakes. Keeps confidences. Shows respect for others; remains firm and fair. Acts with great integrity and character.	Engages in actions inconsistent with words. Blames others for own mistakes. Makes promises that are unrealistic or unkept. Focuses on self-promotion; takes credit for the work and contributions of others. Violates confidences made with others. Gossips or criticizes others behind their back.
Underlying Causes	
Too anxious or timid to deliver unfavorable news. Unable to say "no" at the appropriate time. Unable to maintain a position and follow through. Overly focused on personal ambition and welfare. Avoids conflict. Uncomfortable with how others will respond to the truth.	
Feedback	Get feedback on organization behaviors that demonstrate a high degree of trust. These may include open communication, collaboration, strong innovation, and clear work expectations. Observe your own behavior. Be as objective as possible. Assess if you treat others equitably and fairly—do you have favorites? Get feedback to support your assessment. Contact others outside the unit and find out how to build greater trust, openness, and mutual understanding to achieve common goals. Complete a trust self-assessment. Informal tools are available through online searches.
Study	Observe the behaviors of other leaders who you think are trustworthy. What behaviors do they exhibit that build and maintain trust? Make a list of what they do that you want to model. Study own behaviors. Analyze if own consistency in following up on commitments is less than others. If so, ask or explore why. Learn from mistakes by writing out alternative actions you might have taken. Access the Virtual Improvement Center to complete: Building Working Relationships across Boundaries; Building Trust.
Practice	Let others know what the course of action is and follow through on it. Evaluate personal time available for follow through before making a commitment. Hold a discussion with someone with whom you want to build greater trust and openness.

TAKES DIRECT ACTIONS TO BUILD TRUST

7-17. Building trust is not a passive exercise. Leaders develop trust in their organizations by taking actions that promote trust. Developing others through mentoring, coaching, and counseling are actions that build trust. When a leader mentors effectively, that leader sends a clear message: I trust you to continue the Army profession and build a stronger, more adaptable Army. Leaders build trust by developing positive relationships with peers, superiors, and subordinates (see table 7-10). These leaders do not tolerate misconduct or unfair treatment and they take appropriate action to correct unit dysfunction.

Table 7-10. Takes direct actions to build trust

Strength Indicators	***Need Indicators***
Mentors, coaches, and counsels leaders. Demonstrates care for others. Identifies areas of commonality and builds upon shared experiences. Empowers others in activities and objectives. Unwilling to tolerate discrimination. Corrects actions or attitudes of those who undermine trust. Communicates honestly and openly with others.	Makes little effort to support or develop others. Remains isolated and aloof. Is unwilling to share authority or power in achieving tasks or objectives. Is apathetic towards discrimination, allows distrustful behaviors to persist in unit or team. Is ambiguous, inconsistent, or unclear in communication with others.
UNDERLYING CAUSES	
Does not understand the importance of leader development. Socially anxious, fears failing or appearing weak in front of others. Overly self-focused, focused on own ambitions. General lack of self-confidence in leadership abilities to shape an organization or team. Does not value diversity.	
Feedback	Get feedback from trusted colleagues and mentors on actions they take to build or rebuild trust. Describe the actions taken to build trust in the unit and ask for feedback. Observe the personal actions taken to build trust. Consider how they contribute to building trust. Ask trusted colleagues if these actions had the desired effect. Regularly seek information from others at different organization levels. Find out how clearly orders are communicated through the organization. Seek regular input on your leader development efforts. Assess the extent to which subordinate development occurs. Adjust efforts accordingly. Assess unit morale with command climate surveys or other morale assessments. Allow feedback to be anonymous. Determine whether to take additional actions to build trust.
Study	Observe leaders you think are trustworthy. Consider the actions they take to build or rebuild trust. Effective actions include extending trust to others and planning ways to restore trust. Analyze the trust level in the organization. Consider indications of a breach of trust, such as backstabbing, gossiping, self-serving behavior, verbal abuse, discriminatory behavior, or time spent covering mistakes. Determine actions to remedy and prevent the breaches. Study the unit. Get to know members individually. Understand their strengths, developmental needs, expectations, and motivations. Use this knowledge to establish greater rapport. Study the actions leaders take to rebuild trust if trust has been lost Access the Virtual Improvement Center to complete: Making Influence Count, Rapid Team Stand-up: How to Build Your Team ASAP, Building Working Relationships across Boundaries, or Building Trust. ASAP: as soon as possible
Practice	Clarify task or position expectations. Be clear as to how and when you want to see progress. When developing others through mentoring, coaching, or counseling create agreement on performance change, goals, and specific follow-up or corrective actions. Help subordinates recover from failure by showing understanding and empathy. Counsel subordinates by providing feedback on the course of action, results, and alternatives. If dysfunction or distrustful behaviors occur, take immediate action to correct the behavior. Provide clear feedback about why the actions or attitudes were contributing to a climate of distrust, and describe expectations for the future.

Sustains a Climate of Trust

7-18. A climate of trust requires that the norms and values of the unit create a positive, mutually beneficial environment characterized by openness and risk-tolerance. Leaders sustain this environment by consistently demonstrating these values through their decisions and actions and communicating to others that misconduct will not be tolerated. It is important for leaders to note that setting an example and directing action to build trust are important tools that help to sustain a climate of trust (see table 7-11).

Table 7-11. Sustains a climate of trust

Strength Indicators	***Need Indicators***
Assesses recurring conditions that promote or hinder trust. Keeps people informed of goals, actions, and results. Follows through on actions related to others' expectations. Under-promises and over-delivers. Maintains high unit morale.	Appears insensitive to what promotes or hinders trust. Demonstrates poor communication of goals, actions, and results to others. Shows inconsistency in attitudes or behaviors, does not follow through on actions. Over-promises and under-delivers. Enables poor unit morale.
Underlying Causes	
Overall lack of leadership experience. Insensitivity to the conditions that help create trust or hinder it. General lack of transparency in decisionmaking. Anxiety about perceptions of others, wanting to please. Inability to commit to a particular course of action. Overly ambitious, not focused on the team or causes larger than self.	
Feedback	Regularly meet with key staff to gather feedback on both unit and individual morale, the level of openness in the unit, and factors (positive and negative) which may be influencing trust. Use instruments such as Command Climate Surveys and other assessments to assess unit morale regularly. Low morale is a good indicator of a lack of trust. Encourage frequent informal feedback on unit climate. Note: others will model the values and tone set by a leader. Reward candid, informal feedback. Build trust by acting on the feedback received. If the feedback on climate reveals a weakness in the unit, rebuild trust.
Study	Regularly observe individuals and teams performing their duties during normal operations and trainings in an attempt to gauge the level of trust existing among them. Notice when a climate is distrustful. Study the factors that contributed to the loss of trust. Study the cases of particularly inspiring leaders in both civilian and military culture who created climates of trust in their units, teams, or organizations. Write down the actions they took, and the effect they had on the climate of the organization. Access the Virtual Improvement Center to complete: Making Influence Count; Rapid Team Stand-up: How to Build Your Team ASAP; Building Working Relationships across Boundaries; Building Trust. ASAP: as soon as possible
Practice	Describe unit values surrounding trust frequently. Be clear about how you and all unit members will create a climate of trust. Make building trust an explicit goal. Cultivate risk-tolerance by communicating and demonstrating through actions that taking prudent risks can be appropriate. Create transparency by opening multiple communication channels, including newsletters, reports, and staff meetings to talk openly about performance, mistakes, outcomes, best practices, and resources.

EXTENDS INFLUENCE BEYOND THE CHAIN OF COMMAND

7-19. Leaders can influence beyond their direct line of authority and chain of command. Influence can extend across units, to unified action partners, and to other groups. A key to extending influence beyond the chain of command is creating and communicating a common vision and building agreement. In these situations, leaders use: indirect means of influence, diplomacy, negotiation, mediation, arbitration, partnering, conflict resolution, consensus building, and coordination.

7-20. This competency has two components:

- Understands sphere, means and limits of influence.
- Negotiates, builds consensus and resolves conflict.

UNDERSTANDS SPHERE, MEANS, AND LIMITS OF INFLUENCE

7-21. Leading and influencing others outside the established organizational structure requires specific skills and abilities. Assessing roles of others outside the chain of command, knowing over whom they have authority and influence, and understanding how they are likely to exert that influence is important. By learning about people outside of the chain of command, understanding their interests and viewpoints, and being familiar with internal relationships within the organization, leaders can identify influence techniques likely to work beyond the command chain. Individuals can adjust influence techniques to the situation and parties involved (see table 7-12).

Table 7-12. Understands sphere, means, and limits of influence

Strength Indicators	***Need Indicators***
Assesses situations, missions, and assignments to determine the parties involved in decision making and decision support. Evaluates possible interference or resistance. Reviews organizational structures to understand who reports to whom and informal relationships illustrating who influences whom. Has a good sense of when and when not to influence beyond the chain of command. Gets input from members of own chain of command before influencing others outside it.	Uses the same technique in every situation to influence others. Operates in isolation outside the chain of command when not appropriate. Begins negotiating with others without recognizing their priorities or interests. Relies solely on informal relationships such as colleagues and peers; does not work through the formal command chain. Makes assumptions about others too quickly without getting the facts.
Underlying Causes	
Does not appreciate the potential benefits of understanding spheres of influence. Is impatient; wants to act before understanding relationships. Shields self from criticism or failure; risk averse. Lacks organizational knowledge outside of own chain of command. Is politically insensitive to factors affecting broader Army interests. Is naïve or insensitive to cultural or other differences.	
Feedback	Get feedback on your ability to actively listen, present information so others understand advantages, and be sensitive to the cultural factors in communications. Determine the degree to which you gain cooperation with peers or others outside of the chain of command. Self-assess personal knowledge of another organization, person, or culture. Request feedback on your effectiveness in working with others. For example, ask others about when you effectively demonstrated resilience, patience, confidence, or mental agility.
Study	Learn as much as possible about organization processes and the key players. Gain information about shared common goals between the organization and organizations outside the chain of command; evaluate the similarities and differences. Understand the organization's climate and the origin and reasoning behind key policies, practices, and procedures. Gain insight into the culture, work priorities, and leadership interests outside the chain of command by working on a project or team assignment with another organization. Ask others outside the organization how to gain insight into their organizational priorities. Access the Virtual Improvement Center to complete: Making Influence Count, Building Working Relationships across Boundaries, and The Leader as Follower.

Table 7-12. Understands sphere, means, and limits of influence (continued)

Practice	Practice getting things done using both formal channels and informal networks. Determine who, when, and how to communicate a situation to superiors and the team. Practice face-to-face engagements using role players simulating diverse audiences. Practice explaining the rationale of a tough decision to those affected. Practice focused listening; ask questions to identify points of agreement and contention. Consider alternatives from the viewpoint of others. Ensure team members and subordinates understand the reporting structure in the unit. When communicating decisions or proposing new ideas, clearly articulate the broader strategic benefits to the unit or the Army.

NEGOTIATES, BUILDS CONSENSUS, AND RESOLVES CONFLICT

7-22. The art of persuasion is an important method of extending influence. Proactively involving partners opens communication and helps to work through controversy in a positive and productive way. Building consensus though sharing ideas and seeking common ground helps overcome resistance to an idea or plan (see table 7-13).

Table 7-13. Negotiates, builds consensus, and resolves conflict

Strength Indicators	***Need Indicators***
Identifies individual and group positions and needs. Sees conflict as an opportunity for shared understanding. Facilitates understanding of conflicting positions and possible solutions. Works to collaborate on solving complex problems in ways acceptable to all parties. Builds consensus by ensuring that all team members are heard.	Uses the same technique in every situation to influence others. Negotiates with others without recognizing their priorities or interests. Uses extreme techniques such as being too hard or too soft when resolving conflicts. Isolates team members and pressures them to align with personal goals and priorities. Does not seek to reconcile conflicting positions; only seeks to win. Focuses on negatives of others' interests.
Underlying Causes	
Does not seek the middle ground on issues, but demands that personal identified needs are met. Avoids conflict; uncomfortable in situations that demand identifying the conflict and solving the problem. Is unable or unwilling to look for a common causes or mutual goals. Is uncomfortable or does not like to work with teams towards common goals and priorities. Takes things personally. Does not maintain a solutions-based focus.	
Feedback	Get input from peers about your understanding of negotiation techniques. Ask questions such as "Can you describe a situation I negotiated effectively?" "What could I do to negotiate more effectively?" After presenting a concept or idea to peers, ask for their thoughts and perspectives. Record yourself in a practice session while negotiating a dispute. While viewing the recording, self-assess your actions and note effective and ineffective actions. Before negotiations begin, select several negotiating techniques and practice with a peer to gain insight on technique implementation and the potential drawbacks of each. Request feedback on your skills. Get feedback on your ability to listen actively, to present information so others understand advantages, and your sensitivity to the cultural factors in communication.

Table 7-13. Negotiates, builds consensus, and resolves conflict (continued)

Study	When disputes occur, evaluate areas of common ground between different parties and document findings. List all of the roles and resources that figure in to a goal or priority of the organization. Identify people with whom you may have a common cause or mutual goals. Research the viewpoints of other individuals involved in the negotiation or consensus building. Use those viewpoints accordingly in your argument. Carefully outline personal principles and values so you know when negotiation crosses boundaries. Study the behaviors of strong negotiators or behaviors of successful arbitrators. List specific behaviors they demonstrate that you admire about them.
Practice	Find an opportunity to exercise diplomacy and tact to achieve a favorable outcome. When in a discussion with individuals of differing opinions, practice asking questions likely to result in compromise, such as "What points can we agree upon?" or "What is most important to you and what can you concede? Work to be a team player that can represent personal interests. Anticipate problem areas in complex situations and vary the approach accordingly. Call a team meeting at the first sign that there is tension among group members. Access the Virtual Improvement Center to complete: Extending Influence during Negotiation; Managing Conflict; Building Working Relationships across Boundaries; Building Trust; Navigating Contentious Conversations; Managing Difficult Behavior.

LEADS BY EXAMPLE

7-23. Leaders can influence others by acting in a manner that provides others with an example by which to measure and model their own behavior. Leading by example is a form of influence where leaders provide models rather than explicit direction. Leading by example is a manifestation of character and presence attributes:

- Displays Army Values.
- Displays empathy.
- Exemplifies the Warrior Ethos/Service Ethos.
- Applies discipline.
- Leads with confidence in adverse situations.
- Demonstrates tactical and technical competence.
- Understands the importance of conceptual skills and models them.
- Seeks diverse ideas and points of view.

DISPLAYS ARMY VALUES

7-24. Upon entering the Army, Soldiers learn to uphold a new set of values: the Army Values. The Army Values are a set of principles, standards, and qualities that are essential for Army leaders. The Army recognizes seven values to uphold—loyalty, duty, respect, selfless service, honor, integrity, and personal courage. It is every Soldier's obligation to demonstrate these values through their decisions and actions, and in doing so, set an example for others to follow (see table 7-14 on page 7-16). Demonstrating these values establishes one as a person of character who upholds the Army Ethic in the conduct of mission, performance of duty and all aspects of life.

Table 7-14. Displays Army Values

Strength Indicators	Need Indicators
Displays high standards of duty performance, personal appearance, military and professional bearing, and physical fitness and health. Takes an ethical stance; fosters an ethical climate. Demonstrates good moral judgment and behavior. Completes tasks to standard, on time, and within the commander's intent. Demonstrates determination and persistence when facing adverse situations.	Solves problems using the "easy path" without regard for what is "the right thing to do." Puts personal benefit or comfort ahead of the mission. Hides unpleasant facts that may arouse anger. Is publicly critical of the unit or its leadership, yet does nothing to help.

Underlying Causes
Has not accepted one or more of the Army Values. Overly committed to self-interests, career goals, and personal achievement. Is not able to translate Army Values to personal behaviors. Afraid of facing demands or hardships that following Army Values might bring. Not aware of personal behaviors and how they are perceived by others.

Feedback	Reflect on personal values and the Army Values. If you perceive a conflict, consult a mentor with respected values and judgment for discussion and guidance. Ask co-workers how well they understand the expectations and the standards. Ask peers and subordinates how well they think you uphold the Army Values. How do your behaviors signal your values?
Study	Consider personal behaviors to complete tasks to standard, on time, and within the commander's intent. How do you ensure success of your work? How do you gauge personal adherence to standards? How do you ensure timeliness of completion? Observe other organizational leaders who effectively demonstrate and uphold the Army Values. Tailor the approach to your situation. Analyze the influence of the Army Values on the unit by observing instances and examples of loyalty, duty, respect, selfless service, integrity, honor, and personal courage. What are the consequences when adherence to these values falls short? Study historical military figures who demonstrated determination, persistence, and patience in achieving an objective. What factors led to their success? In times of intense hardship, what actions did they use to overcome adversity? Consider what the Army Values mean and implications for personal behavior and development.
Practice	Exercise initiative by anticipating task requirements before receiving direction. Take responsibility for both yourself and subordinates when an issue arises. Make decisions based on what you know is right. Do not be swayed by circumstances or internal or external factors that may affect the decision. Act according to clear principles rather than the easy path. Foster and encourage an open-door policy with subordinates where they feel comfortable coming to talk to you about ethical and moral challenges they are facing on-duty and how to implement the correct action. Practice what you preach. Demonstrate upholding the Army's Values to others.

Displays Empathy

7-25. Empathy is defined as the ability to share and understand someone else's feelings. The capacity for empathy is an important attribute for leaders to possess. Empathy can allow leaders to understand how their actions will make others feel and react. Empathy can help leaders to understand those that they deal with including other Soldiers, Army Civilians, local populace, and even enemy forces. Being able to see from another's viewpoint enables a leader to understand those around them better (see table 7-15 on page 7-17).

Table 7-15. Displays empathy

Strength Indicators	*Need Indicators*
Reads others' emotional cues. Considers other points of view in decision-making. Reacts appropriately to others' emotional states. Shows compassion when others' are distressed. Predicts how others will react to certain events. Demonstrates ability to establish good rapport.	Shows a lack of concern for others' emotional distress. Displays an inability to take another's perspective. Maintains an egocentric viewpoint in decision-making process. Dehumanizes enemy combatants or local populace.
Underlying Causes	
Problems with or inability to take others' perspectives. Focuses solely on own needs without considering needs of others. Insensitive to emotional cues of others. Failure to identify with other individuals. Overly results focused.	
Feedback	Gather feedback from on your ability to read emotional cues of others. Self-reflect on your successes and failures in perspective taken during the decisionmaking process. Explicitly focus on emotional and social cues in conversations.
Study	Select a role model and study their interactions with others. Read relevant literature on empathy and social perspective taking. Learn more about the pitfalls associated with empathy failures Learn nonverbal cues that can help to indicate a person's emotional state. Access the Virtual Improvement Center to complete: Beyond People Skills: Leveraging Your Understanding of Others module.
Practice	Practice taking perspectives of different people (such as that of a local leader, coalition ally, adversary, or a different military specialty). Imagine what their assumptions and preferences would be. Do this when interacting with a peer or a group. Get to know your subordinates better so you can understand their issues. Use red teaming by taking partner and adversary perspectives to challenge ideas and ensure consideration of all perspectives in the decisionmaking process. Actively combat moral disengagement (convincing oneself that ethical standards do not apply in a certain situation) in peers and subordinates by directly addressing instances when they failed to show concern for others.

EXEMPLIFIES THE WARRIOR ETHOS/SERVICE ETHOS

7-26. The Warrior Ethos and Service Ethos refer to a set of specific professional attitudes and beliefs that characterize the American Soldier and Army Civilian. The Warrior Ethos shapes and guides a leader's actions both on and off the battlefield. Leaders demonstrate the Warrior Ethos or Service Ethos anytime they experience prolonged and demanding conditions that require commitment and resilience to do what is right despite adversity, challenge, and setback (see table 7-16 on page 7-18). For example, tirelessly advocating for a more comprehensive training program on leader development demonstrates the Service Ethos, just as leading others in a combat zone demonstrates the Warrior Ethos.

7-27. While Army Civilians can have a warrior-like ethos, a service ethos fittingly describes the attitudes of Army Civilians who choose to serve the public interest through support and defense of the Constitution. They are committed to the Army and the Constitution and take an oath upon their hiring similar to the oath Soldiers take. In honoring the Service Ethos, Army Civilians help support the needs of the Army and its Soldiers.

Table 7-16. Exemplifies the Warrior Ethos/Service Ethos

Strength Indicators	***Need Indicators***
Removes or fights through obstacles, difficulties, and hardships to accomplish the mission. Demonstrates the will to succeed and perseveres through difficult and complicated situations. Demonstrates physical and emotional courage. Upholds and communicates the Warrior Ethos. Pursues victory over extended periods, regardless of condition.	Gives up when facing a difficult challenge or hardship. Is pessimistic or negative about personal ability to achieve results within organizational constraints. Allows fear of risk to stop action without integrating the risk management process. Hesitates or avoids stepping up when necessary. Demonstrates timidity and hesitation to act.
Underlying Causes	
Failure to internalize the Army Values. Lacks a holistic understanding of the Warrior Ethos and its implications for personal behavior. Exhibits frustration or fatigue from excessively demanding conditions over an extended period. Allows laziness or complacency to compromise the task or mission. Current situation feels hopeless and shows no indication of getting better.	
Feedback	Clarify and understand the scope of new tasks and the relationship to mission accomplishment. Perseverance is valuable when aligned with organizational goals. Request feedback from peers and subordinates on how well you demonstrate determination, persistence, and patience. Determine if patterns exist in how you handle different situations. Ask for feedback from a superior on how well you demonstrate the Warrior Ethos. Identify points where you could have persevered more or where you should have been less persistent to ensure a balance between achieving effective results and wasting time. Request advice from a mentor or trusted advisor before undertaking a difficult task. Have them provide insight into the appropriate steps. Provide as much context as possible and then talk through the situation and possible ways to deal with anticipated difficulties.
Study	Reflect on personal experiences in upholding the Warrior Ethos. In a difficult or prolonged task, what most made you want to give up; what most helped you keep going? If you are having trouble getting something done, reflect on why it is not working and what alternative approaches might succeed. Research historical figures who demonstrated physical and emotional courage and the will to succeed. Read Medal of Honor citations or pick a role model and see how they demonstrate perseverance. What actions and attitudes led them to success? In times of intense hardship, what was their approach to leadership? Identify ways to relieve stress to manage emotional reactions while at work (such as taking deep breaths, counting to ten, or thinking before acting). Study historical figures who demonstrated determination, persistence, and patience in achieving an objective. What factors led to their success? In times of intense hardship, how did they overcome adversity?
Practice	During reviews, consider how the tenets of Warrior Ethos were applied during operations. Volunteer to take the lead on a difficult or prolonged issue. While working through the issue, note the work and progress that occurred toward resolution. When leading, accept responsibility for personal errors and move on. Do not allow criticism of an outcome or setbacks prevent taking the lead or persisting in efforts. When interacting with team members and subordinates, realize resistance and inertia are natural. When they occur, stick to the point, and not take criticism personally.

APPLIES DISCIPLINE

7-28. Discipline is essential for a Soldier and leader. While it is the responsibility of all Soldiers to maintain self-discipline, it is the responsibility of leaders to ensure unit discipline. Self-discipline allows individuals to ensure their behaviors embody the Army Values, make certain that Army standards are met (physical as well as behavioral), and properly accomplish tasks in a timely manner. All of the specified attributes and behaviors of an Army leader are based in self-discipline and the ability to put the needs consistent with support and defense of the Constitution of the United States ahead of one's own needs. Unit discipline encourages a sense of camaraderie, supports a positive climate, and reinforces management systems such as